Hybrid Organic-Inorganic Materials Used to Improve the Environment and Human Health

Hybrid Organic-Inorganic Materials Used to Improve the Environment and Human Health

Editors

Radu Claudiu Fierascu
Florentina Monica Raduly

Basel • Beijing • Wuhan • Barcelona • Belgrade • Novi Sad • Cluj • Manchester

Editors
Radu Claudiu Fierascu
National Institute for Research and
Development in Chemistry and
Petrochemistry
Bucharest
Romania

Florentina Monica Raduly
National Institute for Research and
Development in Chemistry and
Petrochemistry
Bucharest
Romania

Editorial Office
MDPI AG
Grosspeteranlage 5
4052 Basel, Switzerland

This is a reprint of articles from the Special Issue published online in the open access journal *Crystals* (ISSN 2073-4352) (available at: http://www.mdpi.com).

For citation purposes, cite each article independently as indicated on the article page online and as indicated below:

Lastname, A.A.; Lastname, B.B. Article Title. *Journal Name* **Year**, *Volume Number*, Page Range.

ISBN 978-3-7258-1655-2 (Hbk)
ISBN 978-3-7258-1656-9 (PDF)
doi.org/10.3390/books978-3-7258-1656-9

© 2024 by the authors. Articles in this book are Open Access and distributed under the Creative Commons Attribution (CC BY) license. The book as a whole is distributed by MDPI under the terms and conditions of the Creative Commons Attribution-NonCommercial-NoDerivs (CC BY-NC-ND) license.

Contents

About the Editors vii

Preface ix

Florentina Monica Raduly and Radu Claudiu Fierăscu
Hybrid Organic-Inorganic Materials Used to Improve the Environment and Human Health
Reprinted from: *Crystals* **2022**, *12*, 1273, doi:10.3390/cryst12091273 1

Mariya Aleksandrova, Georgi Kolev, Andrey Brigadin and Alexander Lukin
Gas-Sensing Properties of a Carbyne-Enriched Nanocoating Deposited onto Surface Acoustic Wave Composite Substrates with Various Electrode Topologies
Reprinted from: *Crystals* **2022**, *12*, 501, doi:10.3390/cryst12040501 3

Augustin M. Mădălan
Atmospheric Carbon Dioxide Capture as Carbonate into a Luminescent Trinuclear Cd(II) Complex with Tris(2-aminoethyl)amine Tripodal Ligand
Reprinted from: *Crystals* **2021**, *11*, 1480, doi:10.3390/cryst11121480 15

Dalia Santa Cruz-Navarro, Miguel Torres-Rodríguez, Mirella Gutiérrez-Arzaluz, Violeta Mugica-Álvarez and Sibele Berenice Pergher
Comparative Study of Cu/ZSM-5 Catalysts Synthesized by Two Ion-Exchange Methods
Reprinted from: *Crystals* **2022**, *12*, 545, doi:10.3390/cryst12040545 22

Xinlei Ji, Hong Li, Yuan Qin and Jun Yan
Performance Enhancement of Self-Cleaning Cotton Fabric with ZnO NPs and Dicarboxylic Acids
Reprinted from: *Crystals* **2022**, *12*, 214, doi:10.3390/cryst12020214 31

Mirela Galić, Gabrijela Grozdanić, Vladimir Divić and Pavao Marović
Parametric Analyses of the Influence of Temperature, Load Duration, and Interlayer Thickness on a Laminated Glass Structure Exposed to Out-of-Plane Loading
Reprinted from: *Crystals* **2022**, *12*, 838, doi:10.3390/cryst12060838 44

Prabhat Kumar Tripathy and Kunal Mondal
A Molten Salt Electrochemical Process for the Preparation of Cost-Effective p-Block (Coating) Materials
Reprinted from: *Crystals* **2022**, *12*, 385, doi:10.3390/cryst12030385 72

Yangyang Zhang, Na Liu, Haipeng Xie, Jia Liu, Pan Yuan, Junhua Wei, et al.
Modification of $FA_{0.85}MA_{0.15}Pb\ (I_{0.85}Br_{0.15})_3$ Films by NH_2-POSS
Reprinted from: *Crystals* **2021**, *11*, 1544, doi:10.3390/cryst11121544 79

Pollyana P. Firmino, Jaqueline E. Queiroz, Lucas D. Dias, Patricia R. S. Wenceslau, Larissa M. de Souza, Ievgeniia Iermak, et al.
Synthesis, Molecular Structure, Thermal and Spectroscopic Analysis of a Novel Bromochalcone Derivative with Larvicidal Activity
Reprinted from: *Crystals* **2022**, *12*, 440, doi:10.3390/cryst12040440 88

Mehtab Parveen, Mohammad Azeem, Afroz Aslam, Mohammad Azam, Sharmin Siddiqui, Mohammad Tabish, et al.
Isolation, Identification, Spectral Studies and X-ray Crystal Structures of Two Compounds from *Bixa orellana*, DFT Calculations and DNA Binding Studies
Reprinted from: *Crystals* **2022**, *12*, 380, doi:10.3390/cryst12030380 106

Pavel A. Demakov, Alena A. Vasileva, Vladimir A. Lazarenko, Alexey A. Ryadun and Vladimir P. Fedin
Crystal Structures, Thermal and Luminescent Properties of Gadolinium(III) *Trans*-1,4-cyclohexanedicarboxylate Metal-Organic Frameworks
Reprinted from: *Crystals* **2021**, *11*, 1375, doi:10.3390/cryst11111375 . . . . **128**

Florentina Monica Raduly, Valentin Rădiţoiu, Radu Claudiu Fierăscu, Alina Rădiţoiu, Cristian Andi Nicolae and Violeta Purcar
Influence of Organic-Modified Inorganic Matrices on the Optical Properties of Palygorskite–Curcumin-Type Hybrid Materials
Reprinted from: *Crystals* **2022**, *12*, 1005, doi:10.3390/cryst12071005 . . . . **139**

About the Editors

Radu Claudiu Fierascu

Radu Claudiu Fierascu graduated from the University of Bucharest, Faculty of Chemistry, Radiochemistry section, in 2005; he received his MSc in Physical Chemistry and Applied Radiochemistry (2007) and Advanced Materials (2008), and his PhD in Materials Engineering in 2010. Since 2006, he has been an integral part of the National Institute for Research and Development in Chemistry and Petrochemistry—ICECHIM Bucharest. Currently, he holds the positions of senior researcher and Technical Director at ICECHIM. In 2023, he began his four-year term as President of the Scientific Council of ICECHIM. Additionally, since 2019, he has served as a PhD supervisor in the Chemical Engineering field at the National University of Science and Technology Politehnica Bucharest.

Dr. Fierascu has led numerous research, development, and innovation projects as project manager or partner responsible, amassing a total value exceeding 9 million euros. He has authored or co-authored over 170 ISI-indexed papers, with a Hirsch Index of 28 on SCOPUS and over 2200 citations. His contributions include more than 20 books and book chapters, as well as over 30 granted patents and patent applications. In addition to his research and publications, Dr. Fierăscu is an active participant in the academic and scientific community. He serves as a special editor for various Special Issues, is a member of the Editorial Boards of national journals, and is a reviewer for multiple prestigious journals. He has received over 100 awards at various invention and innovation salons.

Dr. Fierascu's scientific research focuses on developing new materials and technologies aimed at improving quality of life, with specific applications in environmental protection, cultural heritage preservation, and biomedical fields.

Florentina Monica Raduly

Florentina Monica Raduly received her Ph.D. in Chemical Engineering from Politehnica University of Bucharest in August 2009. From 2009 to present, she serves as a Research Scientist at the National Institute for Research and Development in Chemistry and Petrochemistry of Bucharest. She directed a research project "New biocompatible products shogaol and curcuminoid-like type used as adjuvantes in cancer radiotherapy" and was a member of more than 25 national and international projects. She has applied for more than 15 national patents. A total of more than 50 ISI research papers with a total of 423 citations in 370 documents (according to SCOPUS) have been published, of which more than 10 high-level papers published by the first author or corresponding author in the fields of materials science, organic-inorganic hybrids, sol–gel processes, coloring and finishing of textile materials, film-forming materials with dyes, antimicrobial delivery systems, including in Gels, Materials, Crystals, Applied Sciences and Coating Journal.

Preface

In recent years, the accelerating pace of climate change has spurred a surge of interest in developing hybrid and nanocomposite materials. These materials are at the forefront of technological advancements, offering solutions that are both environmentally friendly and capable of reducing various pollution factors. Their applications are widespread, influencing sectors such as the automotive, textile, and food industries, and ultimately reflecting on the overall state of human health.

Moreover, the field of medicine has greatly benefited from the application of nanomaterials. The design and engineering of these materials—ranging from metal–organic frameworks, covalent organic frameworks, zeolite materials, and organic–inorganic hybrids, to composites based on graphene, carbon nitride, metals, metal oxides, and polymers—demonstrate significant potential in improving medical treatments and healthcare technologies.

The Special Issue, entitled "**Hybrid Organic–Inorganic Materials Used to Improve the Environment and Human Health**", aims to address one of the most pressing challenges of our time: the improvement of environmental and human health through the innovative application of advanced materials, bringing together cutting-edge research that highlights the pivotal role of hybrid and composite crystalline materials in mitigating pollution and enhancing health outcomes.

The collected works highlight the innovative approaches and practical applications that are instrumental in addressing environmental challenges and improving human health.

We hope that this collection will inspire further research and development in the field of hybrid organic–inorganic materials, fostering new ideas and collaborations that will continue to push the boundaries of what is possible in environmental science and health improvement. We extend our deepest gratitude to the contributors and reviewers whose dedication and expertise have made this Issue possible.

Radu Claudiu Fierascu and Florentina Monica Raduly
Editors

Editorial

Hybrid Organic-Inorganic Materials Used to Improve the Environment and Human Health

Florentina Monica Raduly [1,*] and Radu Claudiu Fierăscu [2,3,*]

1. Laboratory of Functional Dyes and Related Materials, National Institute for Research and Development in Chemistry and Petrochemistry, ICECHIM, 060021 Bucharest, Romania
2. Chemical Technologies Department, Emerging Nanotechnologies Group, National Institute for Research and Development in Chemistry and Petrochemistry, ICECHIM, 060021 Bucharest, Romania
3. Department of Science and Engineering of Oxide Materials and Nanomaterials, University "Politehnica" of Bucharest, 011061 Bucharest, Romania
* Correspondence: monica.raduly@icechim.ro (F.M.R.); fierascu.radu@icechim.ro (R.C.F.)

The Special Issue on "Hybrid Organic-Inorganic Materials Used to Improve the Environment and Human Health" is a collection of 11 original articles (including one communication paper) dedicated to theoretical and experimental research works providing new insights and practical findings in the fields of the environmental protection and human health—related topics.

Almost 26 million tons of plastic waste is generated in Europe every year, according to the specialized studies, most of it being found in the maritime space. Another important pollution factor is food waste. During the process of food decomposition, a significant amount of CO_2 and methane is emitted, which contributes to the potentiating of the greenhouse effect. On the other hand, the excess of nutrients from the land reaches the groundwater, changing the pH of the waters; this favors the excessive growth of algae and bacteria leading to imbalances in the aquatic ecosystem. Thus, various pollution factors compete and request mankind to approach new attitudes, both economically and socially in order to limit the disastrous effects of planet pollution.

In recent years, studies have focused on the development of analysis, monitoring and reduction methods for pollution factors. Thus, hybrid materials were developed (Aleksandrova et al. [1]) that are the basis for the realization of sensors for the detection of volatile organic compounds, while Mădălan [2] synthesized a complex cadmium structure with tris(2-aminoethyl)amine ligand for spontaneous atmospheric CO_2 capture.

The methods to reduce the pollution factors aimed at the catalytic processes, by obtaining new catalysts based on Cu/ZSM-5 (Santa Cruz-Navarro et al. [3]) and the self-cleaning processes that take place after the deposition of ZnO Nps on cotton fabric (Ji et al. [4]).

The use of green energy sources, with direct applications regarding solar energy, is another area that reflects the methods of reducing pollution factors. The materials from which solar cells are made are of real scientific interest and of great importance from an economic perspective. In connection with these, Galić et al. [5] presents a study on the physical properties and mechanical behavior of some sandwich-type structures made of laminated glass with a polymeric intermediary layer. On the other hand, Tripathy et al. [6] and Zhang et al. [7] are concerned with increasing the efficiency of solar cells, by improving the performance characteristics of the components, the films covering the device's surface, respectively.

All these results aimed at reducing the effects of pollution on the environment are directly reflected through the general state of human health. Moreover, the increase in the level of accumulated knowledge and the more intense concern with the methods of protecting the environment have oriented the research on environmentally friendly compounds. Thus, natural compounds for use in fields such as health, industry or agriculture are targeted. The current trend of consuming as many organic products as possible has led to the

Citation: Raduly, F.M.; Fierăscu, R.C. Hybrid Organic-Inorganic Materials Used to Improve the Environment and Human Health. *Crystals* **2022**, *12*, 1273. https://doi.org/10.3390/cryst12091273

Received: 29 August 2022
Accepted: 3 September 2022
Published: 7 September 2022

Publisher's Note: MDPI stays neutral with regard to jurisdictional claims in published maps and institutional affiliations.

Copyright: © 2022 by the authors. Licensee MDPI, Basel, Switzerland. This article is an open access article distributed under the terms and conditions of the Creative Commons Attribution (CC BY) license (https://creativecommons.org/licenses/by/4.0/).

development of natural biocide products (Firmino et al. [8]) and natural food supplements whose bioactive compounds contribute to the treatment or improvement of various ailments (Parveen et al. [9]).

At the same time, the organic–inorganic hybrid materials with luminescent properties based on metal-organic frameworks (MOFs) type structures obtained by Demakov et al. [10] and those based on clays, of the host-dye matrix type synthesized by Raduly et al. [11] find uses in various fields as optical, medical applications and the food industry, which directly influence health and the human condition.

We hope that this collection of papers will meet expectations of readers looking for new advances in the Hybrid Organic-Inorganic Materials Used to Improve the Environment and Human Health field, as well as bringing inspirations for further research work.

Funding: The authors are thankful to the Ministry of Research, Innovation and Digitization, CNCS/CCCDI—UEFISCDI for the financial support of his research projects related to hybrid organic-inorganic materials used to improve the environment and human health topics (e.g., under Grants PN-III-P2-2.1-PED-2019-1471 and Projects to finance excellence in RDI, Contract no. 15 PFE/2021), that to some extent have brought them to being a Guest Editors of this Special Issue.

Acknowledgments: A contribution of all authors is gratefully acknowledged. The authors would like to express his thanks to the Crystals Editorial Office and Technical Coordinator of the Issue for the excellent communication, support, friendly and fully professional attitude.

Conflicts of Interest: The authors declare no conflict of interest.

References

1. Aleksandrova, M.; Kolev, G.; Brigadin, A.; Lukin, A. Gas-Sensing Properties of a Carbyne-Enriched Nanocoating Deposited onto Surface Acoustic Wave Composite Substrates with Various Electrode Topologies. *Crystals* **2022**, *12*, 501. [CrossRef]
2. Mădălan, A.M. Atmospheric Carbon Dioxide Capture as Carbonate into a Luminescent Trinuclear Cd(II) Complex with Tris(2-aminoethyl)amine Tripodal Ligand. *Crystals* **2021**, *11*, 1480. [CrossRef]
3. Santa Cruz-Navarro, D.; Torres-Rodríguez, M.; Gutiérrez-Arzaluz, M.; Mugica-Álvarez, V.; Pergher, S.B. Comparative Study of Cu/ZSM-5 Catalysts Synthesized by Two Ion-Exchange Methods. *Crystals* **2022**, *12*, 545. [CrossRef]
4. Ji, X.; Li, H.; Qin, Y.; Yan, J. Performance Enhancement of Self-Cleaning Cotton Fabric with ZnO NPs and Dicarboxylic Acids. *Crystals* **2022**, *12*, 214. [CrossRef]
5. Galić, M.; Grozdanić, G.; Divić, V.; Marović, P. Parametric Analyses of the Influence of Temperature, Load Duration, and Interlayer Thickness on a Laminated Glass Structure Exposed to Out-of-Plane Loading. *Crystals* **2022**, *12*, 838. [CrossRef]
6. Tripathy, P.K.; Mondal, K. A Molten Salt Electrochemical Process for the Preparation of Cost-Effective p-Block (Coating) Materials. *Crystals* **2022**, *12*, 385. [CrossRef]
7. Zhang, Y.; Liu, N.; Xie, H.; Liu, J.; Yuan, P.; Wei, J.; Zhao, Y.; Yang, B.; Zhang, J.; Wang, S.; et al. Modification of $FA_{0.85}MA_{0.15}Pb(I_{0.85}Br_{0.15})_3$ Films by NH_2-POSS. *Crystals* **2021**, *11*, 1544. [CrossRef]
8. Firmino, P.P.; Queiroz, J.E.; Dias, L.D.; Wenceslau, P.R.S.; de Souza, L.M.; Iermak, I.; Vaz, W.F.; Custódio, J.M.F.; Oliver, A.G.; de Aquino, G.L.B.; et al. Synthesis, Molecular Structure, Thermal and Spectroscopic Analysis of a Novel Bromochalcone Derivative with Larvicidal Activity. *Crystals* **2022**, *12*, 440. [CrossRef]
9. Parveen, M.; Azeem, M.; Aslam, A.; Azam, M.; Siddiqui, S.; Tabish, M.; Malla, A.M.; Min, K.; Rodrigues, V.H.; Al-Resayes, S.I.; et al. Isolation, Identification, Spectral Studies and X-ray Crystal Structures of Two Compounds from *Bixa orellana*, DFT Calculations and DNA Binding Studies. *Crystals* **2022**, *12*, 380. [CrossRef]
10. Demakov, P.A.; Vasileva, A.A.; Lazarenko, V.A.; Ryadun, A.A.; Fedin, V.P. Crystal Structures, Thermal and Luminescent Properties of Gadolinium(III) *Trans*-1,4-cyclohexanedicarboxylate Metal-Organic Frameworks. *Crystals* **2021**, *11*, 1375. [CrossRef]
11. Raduly, F.M.; Rădițoiu, V.; Fierăscu, R.C.; Rădițoiu, A.; Nicolae, C.A.; Purcar, V. Influence of Organic-Modified Inorganic Matrices on the Optical Properties of Palygorskite–Curcumin-Type Hybrid Materials. *Crystals* **2022**, *12*, 1005. [CrossRef]

Article

Gas-Sensing Properties of a Carbyne-Enriched Nanocoating Deposited onto Surface Acoustic Wave Composite Substrates with Various Electrode Topologies

Mariya Aleksandrova [1,*], Georgi Kolev [1], Andrey Brigadin [2] and Alexander Lukin [3]

1. Department of Microelectronics, Technical University of Sofia, 1000 Sofia, Bulgaria; georgi_klv@abv.bg
2. Swissimpianti Sagl, 6828 Balerna, Switzerland; info@swissimpianti.ch
3. Western-Caucasus Research Center, Russian Federation, 352808 Tuapse, Russia; lukin@wcrc.ru
* Correspondence: m_aleksandrova@tu-sofia.bg; Tel.: +359-2-965-30-85

Abstract: The application of carbyne-enriched nanomaterials opens unique possibilities for enhancing the functional properties of several nanomaterials and unlocking their full potential for practical applications in high-end devices. We studied the ethanol-vapor-sensing performance of a carbyne-enriched nanocoating deposited onto surface acoustic wave (SAW) composite substrates with various electrode topologies. The carbyne-enriched nanocoating was grown using the ion-assisted pulse-plasma deposition technique. Such carbon nanostructured metamaterials were named 2D-ordered linear-chain carbon, where they represented a two-dimensionally packed hexagonal array of carbon chains held by the van der Waals forces, with the interchain spacing approximately being between 4.8 and 5.03 Å. The main characteristics of the sensing device, such as dynamic range, linearity, sensitivity, and response and recovery times, were measured as a function of the ethanol concentration. To the authors' knowledge, this was the first time demonstration of the detection ability of carbyne-enriched material to ethanol vapors. The results may pave the path for optimization of these sensor architectures for the precise detection of volatile organic compounds, with applications in the fields of medicine, healthcare, and air composition monitoring.

Keywords: surface acoustic wave composite substrates; microfabrication technology; carbyne-enriched nanocoatings; ion-assisted pulse-plasma deposition; gas-sensing properties

1. Introduction

Following graphene, several new two-dimensional materials have emerged, and their use is expected to have a high impact in our daily life in the near future, including mainstream microelectronics devices [1]. However, it is not only graphene that leads the interest of carbon materials. Other allotropes of carbon exist, including linear molecules, such as polyene (alternating single and triple bonds, i.e., acetylenic carbon) and cumulene (consecutive double bonds). These allotropes are called carbyne. Carbyne contains alternating single and triple carbon bonds, as opposed to polyacetylene, which contains alternating single and double bonds. Carbynes are reported to have a huge Young's modulus of 32.7 TPa, which is 40× that of diamond [2]. Carbynes are thus emerging as a new class of very strong, very tough, and very light material that could be the next revolution of carbon in materials science, fabrics, sensors, electronics, and many more fields. Such new materials are not only important for research, but they will have a huge economic impact as well. For example, the global market for 2D materials is expected to reach USD 400 million by 2025, and this figure is only for the materials and not for the devices making use of those materials [3]. Carbon-based nanomaterials are critical for sensing applications, as they have physical and electronic properties that facilitate the detection of substances in solutions, gaseous compounds, and pollutants through their conductive properties and resonance frequency transmission capacities. However, the use of carbyne films for sensing applications is almost a virgin field with a high potential for innovation and commercial exploitation. Therefore, the fabrication and study of novel carbyne films and modified carbyne films with radical sensing properties and the demonstration of their use as novel chemical sensors would give new insight into the field. To achieve this innovative and wide-spectrum target, different sensor architectures realizing different sensing mechanisms as a result of analyte absorption by the carbyne-enriched layers should be produced and their performance should be compared.

The growing awareness of environmental problems has accelerated research and development in the chemical sensing field and, thus, many high-performing gas and vapor detection techniques have been developed with the potential to be employed in point-of-need implementations [4]. Thanks to these research efforts, several transduction principles have been proposed, implemented, and evaluated in a wide range of applications. When there is a need for highly sensitive, accurate, and rapid detection techniques for chemical agents, an attractive solution for cost and size reduction is to consider the use of micromachined technologies, which offer the capability for on-chip electronics to lower costs and provide high production yields. Among these, cantilevers and capacitive/resistive microsensors have shown huge potential for implementation in our daily life, with some products available on the market already [5,6]. Capacitive sensors were combined with several sensing materials and demonstrated the quantitative and reversible sensing of many gases, such as H_2; O_2; NH_3; NO_2; NH_3; volatile organic compounds (VOCs), such as ethanol, methanol, and formic acid; and relative humidity [7,8]. Recently, a novel gas sensor based upon vapor-induced capacitance modulation of chemically functionalized porous graphene oxide (pGO) was developed [9]. The dielectric pGO matrix was assembled in situ upon an electrode surface through a combined room temperature annealing/freeze-drying process. Extraordinary vapor sensing properties were demonstrated, specifically extremely high sensitivity, wide dynamic range, rapid response and recovery times, fidelity, and detection of a broad range of molecular targets. However, they suffered several drawbacks, such as a long recovery time, low linearity depending on the membrane shape, and their performance was affected by the ambient conditions (temperature, humidity) through the variation of the sensing layer permittivity [10]. In the case of a cantilever-type sensor, there are two key components: a gas sorptive layer, such as carbon-based nanoscale materials and coatings, and a beam transducer. The fundamental resonant frequency of the device depends on the mass loading of the cantilever beam [11]. The beam structures could be designed with a range of dimensions (length, width, and thickness) to explore the effect of the device shape on its resonant frequency and mass sensitivity [12]. The uptake of different gases is monitored as a shift in the device frequency, which is reversible if the gas-sorbing layer interactions are reversible. However, the mechanical motion of these sensors requires specific assembling and packaging processes as compared to static planar devices.

The above-mentioned drawbacks have not been observed for the sensing devices using surface acoustic waves (SAW). They were commercially exploited years ago in industrial applications for the needs of communications, automotive electronics, and environmental sensing [13,14]. Supplying an alternating current (AC) with a certain frequency to electrodes with a specific pattern excites a piezoelectric material, thus generating an acoustic wave that propagates along the surface of the material (surface acoustic wave). When an analyte interacts with this sensing layer physically or chemically, changes occur. Mass and viscosity changes at the sensitive layer can be detected by recording changes in the acoustic wave properties, such as velocity, attenuation, resonant frequency shift, or time delay. Such a structure has been widely investigated for sensing and fluidic applications in advanced lab-on-chip complex devices [15]. Acoustic wave sensors are able to monitor not only mass or density changes but also changes in Young's modulus, viscosity, dielectric, and conductivity properties, wirelessly and in real time [16].

The aim of this study was the exploration and ethanol-sensing performance evaluation of SAW-based devices coated with new carbyne-enriched films grown using the ion-assisted pulse-plasma deposition technique on different electrode topologies. Ethanol is a colorless chemical compound, which is not toxic, but its presence could be an indication of problems with food and drink quality, human medical conditions, industrial manufacturing of raw materials, etc. Therefore, the precise quantitative measurement of ethanol vapors in the range of 40–400 ppm is of great importance for many practical applications. The main characteristics of the sensing device, such as dynamic range, linearity, sensitivity, and response and recovery time were measured as a function of the ethanol concentration. To the authors' knowledge, this is the first demonstration of the detection ability of carbyne-enriched material to ethanol vapors. The results may pave the way for optimization of these sensor architectures for precise detection of volatile organic compounds with applications in the fields of medicine, healthcare, and air composition monitoring.

2. Materials and Methods

For the SAW fabrication, single polished pieces of LiNbO$_3$ wafers, cleaned in ammonia-based solution, were used as substrates. The substrate was SAW grade, which means 128° Y-cut of the LiNbO$_3$ crystal, which was characterized by good temperature stability of the electromechanical coupling factor from 20 °C to 500 °C, where the variations of this factor were <1.8% in this case [17]. This suggested temperature independence of the SAW distribution. Nickel films with a thickness of 120 nm were deposited using vacuum sputtering at a base pressure of 10^{-6} Torr, sputtering voltage of 1.2 kV, and plasma current of 160 mA. A conventional photolithographic patterning procedure with wet anisotropic chemical etching was applied to define the three basic topologies of the samples. The etching solution of HNO$_3$:CH$_3$COOH:H$_2$SO$_4$ = 5:5:2 was prepared. Photos of the prepared samples are shown in Figure 1a–c.

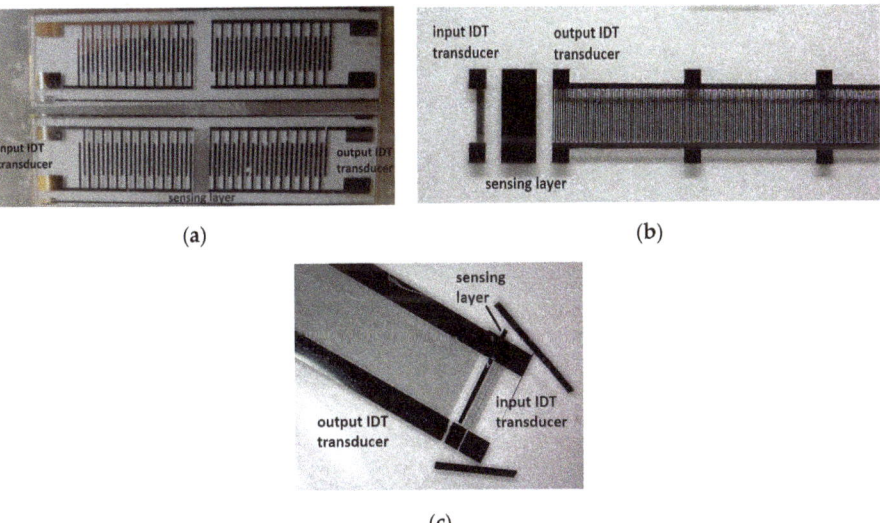

Figure 1. Photos and nomenclature of the prepared SAW structures: (a) device 1, (b) device 2, and (c) device 3.

The geometrical dimensions were as follows: device 1 (Figure 1a)—finger length of the interdigital electrodes (IDT) was 260 µm, finger pitch was 50 µm, and the number of fingers in each transducer from the pair (input and output IDT) was symmetric and equal to 25; device 2 (Figure 1b)—finger length of the interdigital electrodes (IDT) was 150 µm, finger pitch was 25 µm, and the number of fingers in each transducer from the pair (input and output IDT) was asymmetric and equal to 4 and 100, respectively; device 3 (Figure 1c)—finger length of the interdigital electrodes (IDT) was 150 µm, finger pitch was 15 µm, and the number of fingers in each transducer from the pair (input and output IDT) was asymmetric and equal to 4 and 166, respectively. Metal frames and reflectors were additionally formed out of the sensing zone for suppressing wave energy dissipation and improving the signal-to-noise ratio.

Carbyne-enriched nanolayers were selectively grown on the top of the IDT electrodes using the ion-assisted pulse-plasma deposition technique, avoiding the contact pad coating. In previous experiments, new routes to encapsulating oriented linear chains of carbon atoms were established: monatomic carbon filaments in the matrix of amorphous carbon, creating bends, and controlling the end groups in the process of ion-assisted pulse-plasma growth. Within a 2D-ordered linear-chain carbon nanomatrix, the carbon atom wires (CAWs) very weakly interact with each other (due to van der Waals interaction), and therefore, the properties of such nanomatrices are actually determined by the properties of the individual CAWs. The ordered array of the one-dimensional carbon chains packed parallel to one another in hexagonal structures is oriented perpendicular to the substrate surface. The specific energy of the plasma pulse should exceed the breaking energy of the sp^2 bonds (614 kJ/mol) and the sp^3 bonds (348 kJ/mol), but should not exceed the breaking energy of the sp^1 bonds (839 kJ/mol) in the evaporating carbon chains. The controlled bond breaking and sp-phase transformation can be provided through the predictive ion-assisted stimulation with specific energy levels. Raman spectroscopy showed that the line near 2040 cm^{-1} corresponded to the oscillation of long carbon chains with sp^1 hybridization. The obtained spectra revealed multiphonon replicas of the 2040 cm^{-1} line that was evidently associated with the presence of defects in carbon chains [18,19]. Furthermore, the methods of X-ray photoelectron spectroscopy (XPS, Axis Ultra DLD, Kratos, UK) and transmission electron microscopy (Zeiss Leo 912 Omega, Carl Zeiss, Oberkochen, Germany) were previously used, which allowed one to directly register the contribution of sp^1-hybridized carbon bonds, which occur exclusively in chain structures. XPS characterization results demonstrated that the carbyne-like sp^1-hybridized structures were present in the samples. Microscopy images taken showed the discontinuity of the coating on the polycrystalline copper substrate due to adhesion differences, while the monocrystalline silicon allowed for better film quality. A silicon substrate also provides better coupling of vibration modes, leading to stronger chain-related Raman signals in the low-frequency region [20,21]. The thickness of the carbyne-enriched films was ~280 nm for device 1 and device 2 and 140 nm for device 3. The deposition chamber of the experimental setup is shown in Figure 2, illustrating the situation of the target carbon source, ion source, substrate position (Figure 2a), the fixing of the samples (Figure 2b), and the images from the scanning electron microscope (SEM, LYRA I XMU, Tescan, Kohoutovice, Czech Republic) of the carbyne-enriched coatings with the different thicknesses (Figure 2c,d). The deposition conditions were the same for all three samples as follows: distance between the target and the substrates of 1 meter, 3000 pulses, 300 V (the voltage for an arc discharge ignited between the main discharge cathode holding the source of carbon and the main discharge anode holding the substrate), 2000 µF (parameter of the charge of the main capacitor that formed pulses of carbon plasma), 1.5 kV, and 100 mA (parameters of the second argon plasma).

The change in the sensing layer thickness for device 3 was imposed using the attenuation of the surface acoustic wave. The sizes of the SAW devices were calculated, following methodology described elsewhere [22], considering the density and thickness of the coating as related to the IDT electrodes' period and patch, as well as the line length for SAW distribution. The thickness was experimentally confirmed by measuring the magnitude and frequency shift of the output signals as compared to the input signals for all three devices after the sensitive layer growth. They were equal for each combination of IDT's patterns before ethanol vapor absorption and this was the starting point of each measurement.

For the sensing device characterization, a laboratory-made measurement chamber was used. The measurement setup is illustrated in Figure 3.

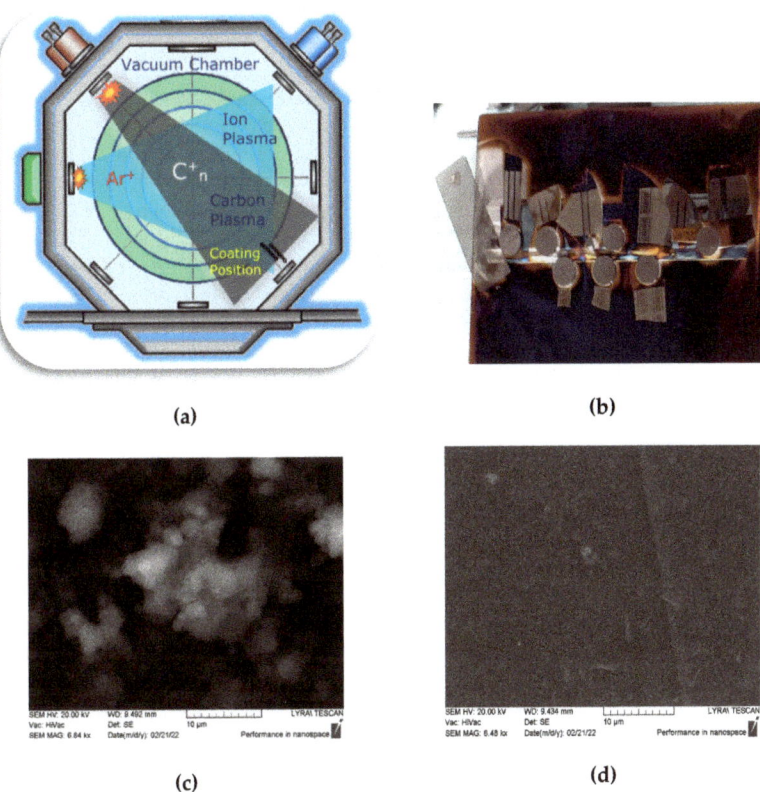

Figure 2. Deposition of carbyne-enriched nanocoatings on the SAW samples: (**a**) deposition chamber, (**b**) fixed samples during the process, (**c**) SEM image of the coating with a thickness of 280 nm, and (**d**) SEM image of the coating with a thickness of 140 nm at the same magnification of 6480×.

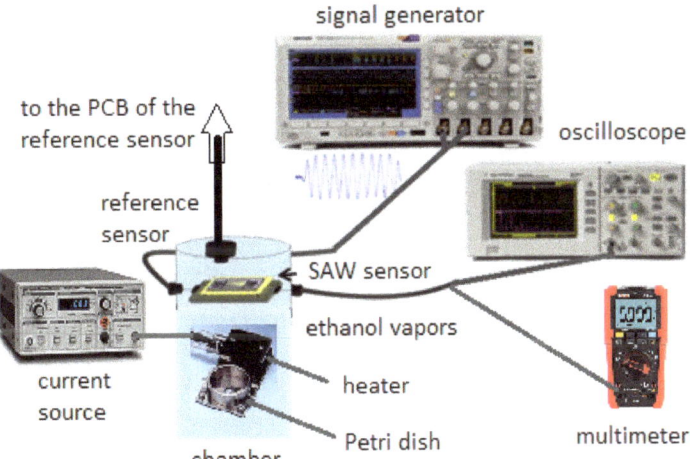

Figure 3. Measurement setup for the functionality testing of the SAW devices with carbyne-enriched nanocoatings and response measurement during ethanol exposure.

Inside a heater was situated so that it was thermally insulated from the chamber. On the top of the heater, there was a Petri dish with a capacity of 3 mL. A current controller with thermocouple feedback set the current through the heater for precise control of the temperature rate change, thus establishing a specific evaporation rate for the 96% concentrated liquid ethanol and converting it into ethanol vapor distributed in the chamber via control of the vaporization time. An additional reference ethanol sensor was mounted close to the device under examination. The SAW structures were first excited with a generator, generating a signal with an amplitude of 20 V from peak to peak and a frequency between 100 kHz and 1 MHz according to the IDT topology (the finest were excited with the highest frequency and vice versa). The output voltage magnitude and shape, as well as the voltage difference and delay time between the output and the input signals, were measured using a Tektronix oscilloscope. The effective output voltage was measured using a multimeter. The control (reference) ethanol sensor was an MQ-3 (ME075) with the necessary resolution.

3. Results and Discussion

As can be seen in Figure 2c, the carbyne-enriched coating was characterized by a fine-grained structure and developed surface, which is favorable for sensing applications and was expected to facilitate vapor absorption.

Figure 4 shows a comparison between the output voltages of all three SAW devices as a function of the stationary conditions in the chamber for each temperature of the heater, resulting in a specific concentration of ethanol vapors in the defined volume of the measurement chamber. One of the Y-axes in each of Figure 4a–c represents the concentration of the ethanol vapors in the closed chamber when ethanol liquid evaporated at a specific temperature. The temperature increased with a slow rate of 0.3 °C/min and could be maintained at an accuracy of 0.01 °C, resulting in a gradual increase in the ethanol concentration with a uniform rate of vaporization of 12 ppm/°C up to 72 °C, where the entire quantity of ethanol was fully vaporized. This setup provided a linear relation of the ethanol vapor concentration with respect to the temperature that was used to estimate the linearity of the response of the SAW devices.

As can be seen in Figure 4a, the output voltage of the sample with the largest IDTs displayed atypical behavior, as the number of absorbed ethanol molecules caused an increase in the voltage, which suggested an increase in the intensity of the surface acoustic wave. This was possible only if the device fell into a new resonance mode due to the mass loading with ethanol vapors that provoked an up to 5 times greater amplitude of the output voltage. However, this phenomenon was observed in a limited concentration range and could not be applied to the whole range of the measured concentrations. Although the exact reasons for this electromechanical behavior are still not fully clarified, device 1 needs attention because of the promising sensitivity of 270 µV/ppm, which was the highest out of the three kinds of samples. The dynamic range of this device was found to be ~50–550 ppm and the deviation from the linearity in this range was determined to be 0.94%. As this was the narrowest dynamic range from the studied devices, it might be considered that the topology of the sample, which was characterized by the lowest number of IDTs and the greatest patch distance in between the neighboring electrode fingers, led to the lowest excitation frequency of 100 kHz. The lowest operational frequency, in combination with the greatest interelectrode gap coated with carbyne-enriched materials, probably resulted in the formation of multiple reactive capacitive components along the length of the SAW line that could strongly affect the resonant frequency change of the device and could partially explain the atypical direction of change of the output voltage. This was due to the accumulated charges on the microcapacitors that were formed by the multiplying effect at the end of the sensing line.

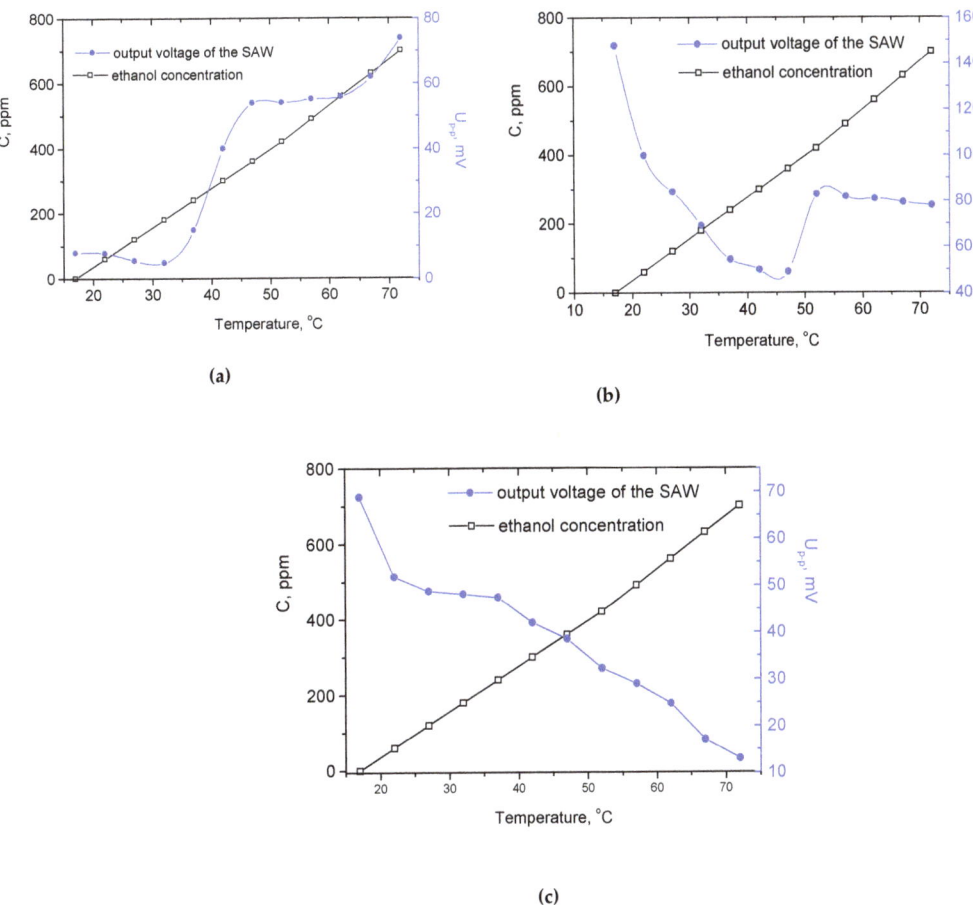

Figure 4. Peak-to-peak voltage of the sensors' outputs at different temperatures in the chamber, corresponding to different evaporated concentrations of ethanol in its volume: (**a**) device 1, (**b**) device 2, and (**c**) device 3.

As can be seen in Figure 4b, device 2 showed typical behavior of attenuation regarding the output voltage, and thus, the energy of the acoustic wave with the increase of the concentration of the absorbed ethanol molecules. Following the assumptions for the previous device 1, it can be concluded that the IDT electrode gaps were sufficiently small to not affect the distribution of the energy conversion by the piezoelectric substrate. Despite the slightly lower sensitivity of device 2 of 221 µV/ppm as compared to device 1, the dynamic range was ~330 ppm (30–660 ppm), which was a broader range with a linear response. A non-linearity of ~3.9% was obtained for this device. Figure 4c shows the voltage response characteristic for the sensing device 3, which was characterized by smaller IDT features than device 1 and device 2 but also had a carbyne-enriched nanolayer with half the thickness to avoid attenuation of the SAW. It exhibited the lowest sensitivity out of all studied devices of ~80 µV/ppm, which confirmed the assumption for the van der Waals adsorption sensing mechanism. In this case, the sensitivity was limited by the thinner sensing film, which limited the adsorption capacity of the device. The non-linearity of the device was found to be 0.97%, which was close to device 1, but in contrast to this sensor, device 3 was characterized by the broadest dynamic range of ~428 ppm (59–487 ppm) and an attenuation of the output signal with the ethanol mass loading effect increase.

The response time was determined using the well-known methodology for pulsed signals, where the time for establishing 10% to 90% from a certain equilibrium condition is called the rise time and the time for the switching of the equilibrium conditions in the opposite direction is called the fall time (see Figure 5a). For device 1, a response time (rise time) of 28 s and a recovery time (fall time) of 20.1 s were measured for an ethanol concentration of C = 400 ppm, corresponding to the middle zone of linear response characteristic. This can be classified as a slow response time. The time response characteristics were relatively symmetric regarding the rise and fall times. For device 2 (Figure 5b), a response time of 20 s and a recovery time of 14.1 s were measured for the middle zone of linear response. A slightly faster response time was measured than for device 1, and a faster recovery time with a small difference between both times. For device 3 (Figure 5c), 31 s was measured for the response time and 15 s for the recovery time in the middle zone of the linear response. This was the longest response time of all the tested samples and there was a large difference between both times.

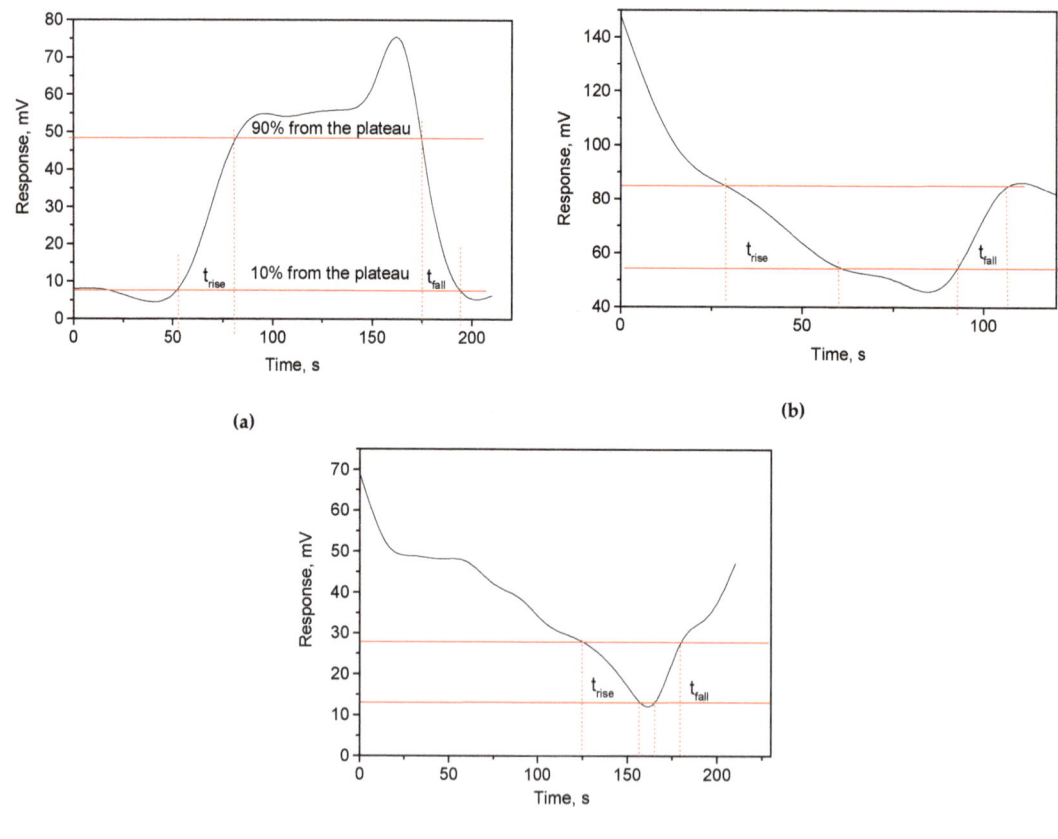

Figure 5. Response and recovery times for the ethanol concentration selected as a middle point of the linear zone of the sensors' responses: (**a**) device 1, (**b**) device 2, and (**c**) device 3.

All three devices exhibited relatively good recovery ability after unloading from the ethanol molecules (Figure 5a–c). The unloading effect was achieved by a simple blowing of the chamber with an inert gas without additional heating of the sensing components. It was suggested that the van der Waals interaction between the vapor molecules and the carbyne-enriched film was boosted by the highly developed surface of the coating, as shown in Figure 2c. Thus, the higher sensitivity of devices 1 and 2, as compared with device 3, was ascribed to their higher thickness of the sensing layer, which was related to a higher capacity for physical adsorption. Therefore, full recovery of the initial state of each device can be expected if post-measurement heating of the samples is applied.

Figure 6a–c shows the changes in the time delay and voltage attenuation between the output and input signal of the IDT electrodes. These parameters determine the resonance frequency shift and damping as a function of the film mass or density shifts due to the ethanol concentration. Their behavior was distinct for each of the electrode topologies previously outlined: (a) For device 1, the time delay decreased linearly with the concentration increase from zero to 100 ppm and later from 250 to 320 ppm, while the damping or voltage difference was nearly constant, with the exception of the 430–440 ppm, where there was a maximum corresponding to strong and stable detection. (b) For device 2, the voltage difference changed more rapidly several times with the concentration with different magnitudes, while the time delay increased from the unloaded resonator value to the fully loaded value and was established at a constant level. This could mean that the upper surface of the film uniformly absorbed the vapor pressure and led to a larger deformation in the film, causing the instability of the signal. (c) For device 3, the two measured quantities increased relatively rapidly and were stable with the concentration. Some unexpected variations around 280–450 ppm that occurred in the time delay and voltage attenuation could be attributed to the changes in the mode of wave propagation due to the device's fine geometry.

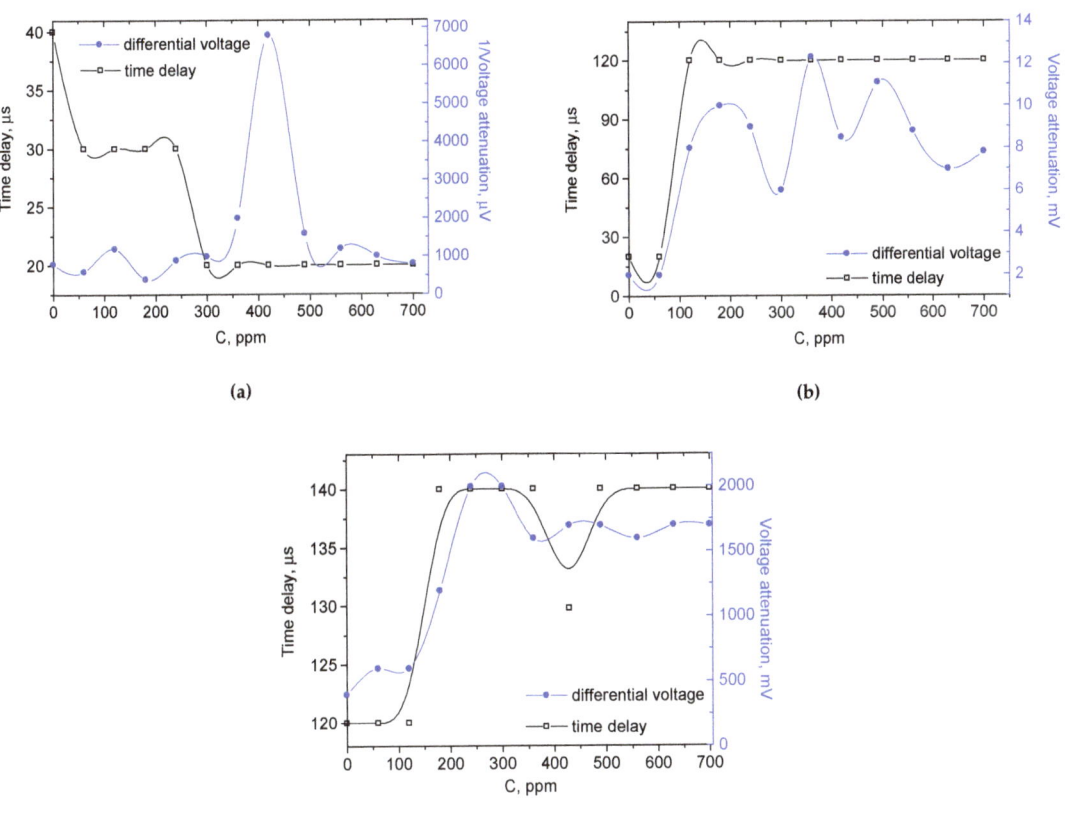

Figure 6. Voltage differences between the output and input signals of the SAW and time delay of the signal vs. ethanol concentration: (**a**) device 1, (**b**) device 2, and (**c**) device 3.

Figure 7a–c shows the oscillograms of the output voltage to demonstrate the lack of signal distortions that might affect the measurements. The total harmonic distortions (THDs) were calculated for the cases of interest and it was found that for most of the cases, they are less than 1%, except for device 1 for the cases after ethanol loading with a concentration of 200 and 300 ppm, where the THDs were 1.41% and 1.7%, respectively. Therefore, the reflectors designed according to the used IDTs topologies seemed to be suitable for suppression of the possible induced noise. The noise-susceptible structures exhibited significant stability to random effects that could induce an atypically large variation. This could be observed as a smooth output signal, not only for the devices before loading but also for the concentrations of ethanol close to the middle linear zone of detection for each topology.

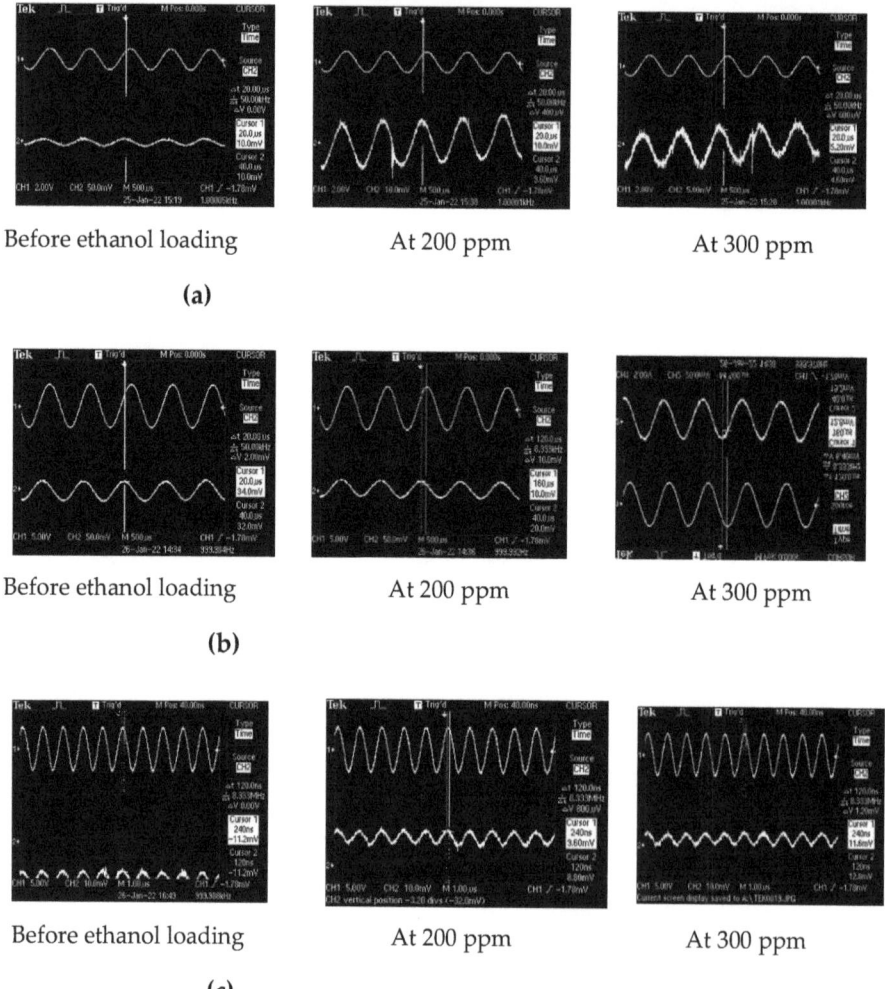

Figure 7. SAW's output signal shapes and distortions at different ethanol concentrations: (**a**) device 1, (**b**) device 2, and (**c**) device 3.

Summarization of the values of the basing sensing parameters and characteristics for the three studied devices can be found in the comparative Table 1.

Table 1. Main ethanol detection parameters of SAW devices with different IDT patterns and using carbyne-enriched coating as a sensing material.

Parameter/ Device No.	Geometry and Sensing Film Thickness, nm	Sensitivity, µV/ppm	Dynamic Range, ppm	Linearity, %	Response Time, s	Recovery Time, s
Device 1	Coarse combs/280 nm	227	184	0.94	28	20.1
Device 2	Fine combs/280 nm	221	330	3.9	20	14.1
Device 3	Fine combs/140 nm	80	428	0.97	31	15

4. Conclusions

The prepared sensing structures exhibited strong dependences on the detection parameters of the electrodes' topologies. It was found that the finer comb geometry was preferable for the broad dynamic range and linear response of the device. One of the benefits related to this principle of ethanol measurement was that the optimal operating temperature of such sensors was preferably low, thus having a low power consumption.

Although the dominant interaction mechanism between the organic analyte and the carbyne-enriched coating seemed to be physical adsorption, full recovery of the samples' initial state after exposure to certain vapor concentrations was not observed. Therefore, future optimization of this parameter can be achieved via the integration of a heating component on the backside of the piezoelectric substrate, facilitating the desorption process.

Due to the most promising results for sensing device 3, it is worth further investigation and optimization. Future studies will be focused on the exposure of the devices to a variety of volatile organic compounds and humidity in order to investigate their selectivity. Moreover, future experiments will be related to full restoration of the initial condition of the sensor, which is possible with full desorption of the ethanol molecules. For this purpose, the integration of a heater on the backside of the substrate is planned and the effect of the temperature on the sensor's full restoration (unloading) will be studied.

Author Contributions: Conceptualization, M.A. and A.L.; methodology, M.A. and A.L.; validation, A.B. and G.K.; formal analysis, M.A., G.K., and A.L.; investigation, M.A., A.B., G.K., and A.L.; resources, M.A. and A.B.; writing—original draft preparation, all authors; writing—review and editing, all authors; visualization, M.A.; supervision, M.A. and A.L.; project administration, M.A. and A.L.; funding acquisition, M.A. and A.L. All authors have read and agreed to the published version of the manuscript.

Funding: This research was funded by the Bulgarian National Science Fund, grant number KP-06-DO2/2 form 2020, under the ERA.NET RUS+ project. The research work of the Russian research team was jointly supported and funded by Russian Foundation for Basic Research (RFBR) according to the research project No. 20-58-46014.

Institutional Review Board Statement: Not applicable.

Informed Consent Statement: Not applicable.

Data Availability Statement: Data are available on request due to potential proprietary restrictions.

Conflicts of Interest: The authors declare no conflict of interest. The funders had no role in the design of the study; in the collection, analyses, or interpretation of data; in the writing of the manuscript; or in the decision to publish the results.

References

1. Flood, P.; Babaev, V.; Khvostov, V.; Novikov, N.; Guseva, M. Carbon Material with a Highly Ordered Linear-Chain Structure. In *Polyynes Synthesis, Properties and Applications*; Cataldo, F., Ed.; Taylor & Francis Group: Didcot, UK, 2005; pp. 219–252. [CrossRef]
2. Aleksandrov, A.F.; Guseva, M.B.; Savchenko, N.V.; Streletskii, O.A.; Khvostov, V.V. The Film of Two-Dimensionally Ordered Linear-Chain Carbon and Method of Its Production. Russian Patent Application No. 2564288, 10 May 2015.
3. Graphene-Info Updates All Its Graphene Market Report. Available online: https://www.graphene-info.com/graphene-info-updates-all-its-graphene-market-report-6 (accessed on 1 March 2022).

4. Akinwande, D.; Huyghebaert, C.; Wang, C.-H.; Serna, M.I.; Goossens, S.; Li, L.-J.; Wong, H.-S.P.; Koppens, F.H.L. Graphene and two-dimensional materials for silicon technology. *Nature* **2019**, *573*, 507–518. [CrossRef] [PubMed]
5. Grogan, C.; Amarandei, G.; Lawless, S.; Pedreschi, F.; Lyng, F.; Benito-Lopez, F.; Raiteri, R.; Florea, L. Silicon Microcantilever Sensors to Detect the Reversible Conformational Change of a Molecular Switch, Spiropyan. *Sensors* **2020**, *20*, 854. [CrossRef] [PubMed]
6. Boom, B.A.; Bertolini, A.; Hennes, E.; van den Brand, J.F.J. Gas Damping in Capacitive MEMS Transducers in the Free Molecular Flow Regime. *Sensors* **2021**, *21*, 2566. [CrossRef] [PubMed]
7. Yao, X.; Zhang, Y.; Cui, Y. A Microfabricated Transparent Capacitor for Sensing Ethanol. *J. Microelectromechanical Syst.* **2019**, *28*, 164–169. [CrossRef]
8. Yang, M.Z.; Dai, C.L. A Capacitive Ammonia Sensor Using the Commercial 0.18 μm CMOS Process. In *Advanced Materials Research*; Trans Tech Publications Ltd.: Kapellweg, Switzerland, 2015.
9. Teradal, N.L.; Marx, S.; Morag, A.; Jelinek, R. Porous graphene oxide chemi-capacitor vapor sensor array. *J. Mater. Chem. C* **2017**, *5*, 1128–1135. [CrossRef]
10. Mohammed, Z.; Elfadel, I.A.M.; Rasras, M. Monolithic Multi Degree of Freedom (MDoF) Capacitive MEMS Accelerometers. *Micromachines* **2018**, *9*, 602. [CrossRef] [PubMed]
11. Mouro, J.; Pinto, R.; Paoletti, P.; Tiribilli, B. Microcantilever: Dynamical Response for Mass Sensing and Fluid Characterization. *Sensors* **2021**, *21*, 115. [CrossRef] [PubMed]
12. Bouchaala, A.; Nayfeh, A.H.; Younis, M.I. Frequency Shifts of Micro and Nano Cantilever Beam Resonators Due to Added Masses. *J. Dyn. Sys. Meas. Control.* **2016**, *138*, 091002. [CrossRef]
13. Lu, X.; Cui, M.; Yi, Q.; Kamrani, A. Detection of mutant genes with different types of biosensor methods. *TrAC Trends Anal. Chem.* **2020**, *126*, 115860. [CrossRef]
14. Yeo, L.Y.; Friend, J.R. Surface Acoustic Wave Microfluidics. *Annu. Rev. Fluid Mech.* **2014**, *46*, 379–406. [CrossRef]
15. Chen, Q.; Wang, D.; Cai, G.; Xiong, Y.; Li, Y.; Wang, M.; Huo, H.; Lin, J. Fast and sensitive detection of foodborne pathogen using electrochemical impedance analysis, urease catalysis and microfluidics. *Biosens. Bioelectron.* **2016**, *86*, 770–776. [CrossRef] [PubMed]
16. Agostini, M.; Greco, G.; Cecchini, M. Full-SAW Microfluidics-Based Lab-on-a-Chip for Biosensing. *IEEE Access* **2019**, *7*, 70901–70909. [CrossRef]
17. Chen, F.; Kong, L.; Song, W.; Jiang, C.; Tian, S.; Yu, F.; Qin, L.; Wang, C.; Zhao, X. The electromechanical features of $LiNbO_3$ crystal for potential high temperature piezoelectric applications. *J. Mater.* **2019**, *5*, 73–80. [CrossRef]
18. Streletskiy, O.A.; Zavidovskiy, I.A.; Nischak, O.Y.; Pavlikov, A.V. Multiphonon replicas in Raman spectra and conductivity properties of carbon films with different concentrations of sp1-bonds. *Thin Solid Film.* **2019**, *671*, 31–35. [CrossRef]
19. Buntov, E.A.; Zatsepin, A.F.; Guseva, M.B.; Ponosov, Y.S. 2D-ordered kinked carbyne chains: DFT modeling and Raman characterization. *Carbon* **2017**, *117*, 271–278. [CrossRef]
20. Streletskiy, O.A.; Zavidovskiy, I.A.; Nischak, O.Y.; Dvoryak, S.V. Electrical conductivity and structural properties of a-C:N films deposited by ion-assisted pulse-arc sputtering. *Thin Solid Film.* **2020**, *701*, 137948. [CrossRef]
21. Boqizoda, D.; Zatsepin, A.; Buntov, E.; Slesarev, A.; Osheva, D.; Kitayeva, T. Macroscopic behavior and microscopic factors of electron emission from chained nanocarbon coatings. *C* **2019**, *5*, 55. [CrossRef]
22. Ballantine, D.S.; White, R.; Martin, S.; Ricco, J.; Zellers, E.; Frye, G.; Wohltjen, H. *Acoustic Wave Sensors Theory, Design and Physico-Chemical Applications: A Volume in Applications of Modern Acoustics*; Elsevier Inc.: Amsterdam, The Netherlands, 1997.

Communication

Atmospheric Carbon Dioxide Capture as Carbonate into a Luminescent Trinuclear Cd(II) Complex with Tris(2-aminoethyl)amine Tripodal Ligand

Augustin M. Mădălan

Inorganic Chemistry Department, Faculty of Chemistry, University of Bucharest, 23 Dumbrava Rosie, 020464 Bucharest, Romania; augustin.madalan@chimie.unibuc.ro

Abstract: Spontaneous atmospheric CO_2 capture as carbonate anion occurred in the synthesis of a trinuclear Cd(II) complex with tris(2-aminoethyl)amine ligand. In reaction two types of compounds were obtained and structurally characterized by X-ray diffraction on a single crystal: initially [{Cd(tren)}$_3$(tren)](ClO$_4$)$_6$·2H$_2$O (**1**) and subsequently [{Cd(tren)}$_3$(tren)][{Cd(tren)}$_3$(μ_3-ηCO$_3$)](ClO$_4$)$_{10}$ (**2**). The carbonate anion replaces partially the bridging tren molecule and coordinates in a μ_3 fashion. The luminescent properties of the compounds were investigated.

Keywords: cadmium complexes; tripodal ligands; CO_2 capture; luminescence

Citation: Mădălan, A.M. Atmospheric Carbon Dioxide Capture as Carbonate into a Luminescent Trinuclear Cd(II) Complex with Tris(2-aminoethyl)amine Tripodal Ligand. *Crystals* **2021**, *11*, 1480. https://doi.org/10.3390/cryst11121480

Academic Editors: Radu Claudiu Fierascu, Florentina Monica Raduly and Shujun Zhang

Received: 31 October 2021
Accepted: 26 November 2021
Published: 29 November 2021

Publisher's Note: MDPI stays neutral with regard to jurisdictional claims in published maps and institutional affiliations.

Copyright: © 2021 by the author. Licensee MDPI, Basel, Switzerland. This article is an open access article distributed under the terms and conditions of the Creative Commons Attribution (CC BY) license (https://creativecommons.org/licenses/by/4.0/).

1. Introduction

Chemical fixation of CO_2 by metal complexes, especially as carbonate anion, attracted particular attention in terms of fundamental research but also of potential applicability for reducing concentration of this major greenhouse gas. Carbonate anion is a versatile bridging ligand, able to link multiple metal ions in various coordination modes [1–3]. Among these numerous coordination modes of the carbonate anion, several types of μ_3 bridges were reported: *syn-syn* μ_3 [4], *syn-anti* μ_3 [5–8], $\mu_3-\kappa^2:\kappa^2:\kappa^2$ [7,9–12], $\mu_3-\kappa^2:\kappa^2:\eta^1$ [13,14], and $\mu_3-\kappa^2:\eta^1:\eta^1$ [7,15,16] (Scheme 1).

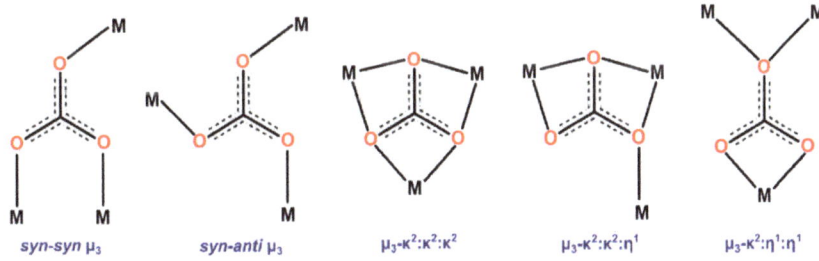

Scheme 1. Several types of μ_3-carbonato bridges.

Most of the carbonate metal complexes are obtained by the addition of carbonate or bicarbonate salts to the reaction mixtures. Atmospheric CO_2 capture is less frequent and it is usually helped by the presence of basic reagents. In this approach, we intended synthesis of trinuclear Cd(II) complexes with tris(2-aminoethyl)amine ligand (tren), in which the tren ligand must play a double role: to act as chelating tripodal ligand capping the metal ions and, also, as bridging ligand connecting three capped Cd(II) ions. This dual behavior of the tren ligand was already reported in the presence of Cu(II) ions, generating trinuclear complexes [17–19], and Cd(II) ions, leading to heptanuclear complexes [20].

In the trinuclear complexes, the bridging tren ligand is replaceable by carbonate anion because it pre-organizes the metal ions in a trigonal manner favorable to a μ_3-$\kappa^2:\kappa^2:\kappa^2$

coordination mode, and the basic free ligand enhances the atmospheric CO_2 capture. As potential mechanisms for carbonate formation, we can mention here hydration of the CO_2 followed by the deprotonation of the carbonic acid in basic condition [2], or generation of tren carbamates followed by hydrolysis and deprotonation [21]. Despite their toxicity, the interest in cadmium complexes is justified by their potential luminescent properties, in the attempt to combine the CO_2 capture ability with sensing application.

2. Materials and Methods

2.1. Synthesis

The chemicals and the solvents were acquired from commercial sources and used as received.

Syntheses of $[\{Cd(tren)\}_3(tren)](ClO_4)_6 \cdot 2H_2O$ (**1**) and $[\{Cd(tren)\}_3(tren)][\{Cd(tren)\}_3(\mu_3\text{-}CO_3)](ClO_4)_{10}$ (**2**).

An amount of 0.8 mmol of tris(2-aminoethyl)amine and 0.2 mmol of $Cd(ClO_4)_2 \cdot 6H_2O$ were dissolved in 30 mL of water and 15 mL ethanol. The mixture was stirred for 30 min and then filtered. The solution was left for slow evaporation at room temperature. Colorless crystals of $[\{Cd(tren)\}_3(tren)](ClO_4)_6 \cdot 2H_2O$ (**1**) appeared after one to two days on the wall of the beaker, while the colorless crystals of $[\{Cd(tren)\}_3(tren)][\{Cd(tren)\}_3(\mu_3\text{-}CO_3)](ClO_4)_{10}$ (**2**) were obtained after one week, preponderantly on the bottom of the beaker. The crystals were separated mechanically (yields: **1** about 20%, **2** about 45%). Compound **2** was obtained as the sole crystalline product by slow evaporation of the solution containing tren and cadmium perchlorate in a 4:1 molar ratio which resulted after bubbling CO_2 in solution for 10 min. The bubbling of CO_2 in solution was used to increase the amount of CO_2 physically dissolved in the water-ethanol mixture. The crystals were collected prior to total evaporation of the solvent in order to avoid contamination (the yield was 60%).

FT-IR (cm^{-1}) **1**: 3603 br, 3348 s, 3295 s, 3174 w, 2956 m, 2905 m, 1590 m, 1465 w, 1316 w, 1055 vs, 977 s, 880 m, 614 s. FT-IR (cm^{-1}) **2**: 3556 w, 3352 s, 3294 s, 3173 w, 2958 w, 2855 w, 1590 m, 1464 m, 1316 w, 1061vs, 983 s, 614 s.

Elemental analysis: (**1**) Calculated: C, 18.54; H, 4.93; N, 14.41%. Found: C, 18.91; H, 4.52; N, 14.07%; (**2**) Calculated: C, 18.76; H, 4.61; N, 14.25%. Found: C, 18.96; H, 4.37; N, 13.86%.

2.2. Physical Measurements

2.2.1. X-ray Structure Determination

X-ray single crystal diffraction measurements were made on a Rigaku XtaLAB Synergy-S diffractometer operating with a Mo-Kα (λ = 0.71073 Å) micro-focus sealed X-ray tube. The structures were solved by direct methods and refined by full-matrix least squares techniques based on F^2. The non-H atoms were refined with anisotropic displacement parameters. Calculations were performed using the SHELX-2018 crystallographic software package. A summary of the crystallographic data and the structure refinement for crystals **1** and **2** are presented in Table 1. The CCDC reference numbers are: 2119345 and 2119346.

2.2.2. Spectroscopy

IR spectra were collected on an FT-IR Bruker Vertex 70 spectrometer (Billerica, MA, USA) in the 4500–400 cm^{-1} range using the ATR technique. The abbreviations used are: w = weak, m = medium, s = strong, v = very, br = broad. Absorption spectra on powder (diffuse reflectance technique) were measured with a JASCO V-670 (Oklahoma, OK, USA) spectrophotometer. The fluorescence spectra were recorded on powder using a JASCO FP-6500 (Oklahoma, OK, USA) spectrofluorometer.

Table 1. Crystallographic data, details of data collection and structure refinement parameters for compounds **1** and **2**.

Compound	1	2
Chemical formula	$C_{24}H_{76}Cd_3Cl_6N_{16}O_{26}$	$C_{43}H_{126}Cd_6Cl_{10}N_{28}O_{43}$
M (g mol^{-1})	1554.90	2752.61
Temperature, (K)	293 (2)	293 (2)
Wavelength, (Å)	0.71073	0.71073
Crystal system	Triclinic	Monoclinic
Space group	P-1	P21
a (Å)	14.8621 (7)	17.4973 (6)
b (Å)	15.3643 (6)	12.2107 (4)
c (Å)	15.5852 (8)	23.5653 (8)
α (°)	99.667 (4)	90
β (°)	96.885 (4)	100.059 (3)
γ (°)	116.107 (4)	90
V (Å^3)	3073.9 (3)	4957.4 (3)
Z	2	2
D_c (g cm^{-3})	1.680	1.844
μ (mm^{-1})	1.372	1.629
F (000)	1572	2764
Goodness-of-fit on F^2	1.023	1.046
Final R_1, wR_2 [I > 2σ(I)]	0.0806, 0.2320	0.0523, 0.1404
R_1, wR_2 (all data)	0.0960, 0.2529	0.0598, 0.1473

3. Results

By reacting tris(2-aminoethyl) amine with cadmium perchlorate in 4:1 molar ratio, two different compounds were obtained: initially [{Cd(tren)}$_3$(tren)](ClO$_4$)$_6$·2H$_2$O (**1**) and subsequently [{Cd(tren)}$_3$(tren)][{Cd(tren)}$_3$(μ$_3$-CO$_3$)](ClO$_4$)$_{10}$ (**2**). Both compounds contain trinuclear Cd (II) complexes.

3.1. Description of the Crystal Structures

Compound **1** crystallizes in the *triclinic P-1* space group and the asymmetric unit consists of trinuclear [{Cd(tren)}$_3$(tren)]$^{6+}$ cations, six perchlorate anions and two crystallization water molecules (Figure 1a). The trinuclear cationic complexes present a tripodal shape (Figure 1b). The Cd(II) ions are pentacoordinated with a trigonal bipyramidal stereochemistry. The equatorial positions and one axial position are occupied by a tren molecule coordinating as a tetradentate chelating ligand. Three such units are bridged by the fourth tren molecule coordinating through the primary amine groups in the free axial position of the Cd(II) ions. One perchlorate ion is hosted between the arms of the tripodal cation.

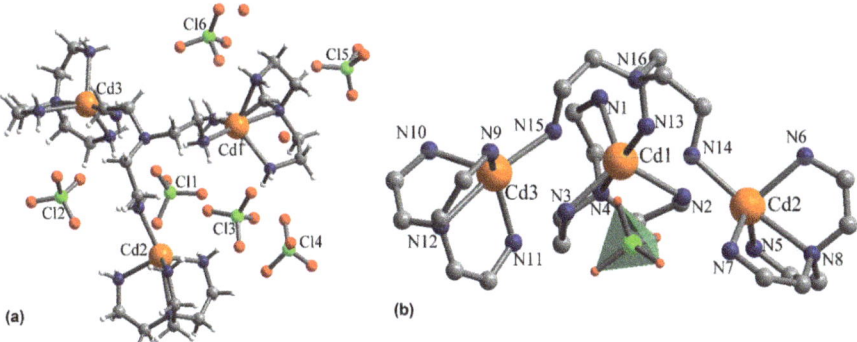

Figure 1. (**a**) View of the asymmetric unit in crystal **1** and (**b**) perspective view of the tripodal trinuclear cation [{Cd(tren)}$_3$(tren)]$^{6+}$ with the perchlorate anion hosted inside (the hydrogen atoms were omitted for clarity).

The equatorial Cd-N bond lengths are: Cd1-N1 = 2.299(7), Cd1-N2 = 2.294(7), Cd1-N3 = 2.277(8), Cd2-N5 = 2.258(8), Cd2-N6 = 2.300(8), Cd2-N7 = 2.310(9), Cd3-N9 = 2.267(10), Cd3-N10 = 2.288(13) and Cd3-N11 = 2.264(14) Å, while the axial Cd-N bond lengths are: Cd1-N4 = 2.430(6), Cd1-N13 = 2.252(6), Cd2-N8 = 2.459(7), Cd2-N14 = 2.282(6), Cd3-N12 = 2.418(8) and Cd3-N15 = 2.282(8) Å. The Cd···Cd distances within the trinuclear complex are: Cd1···Cd2 = 6.67, Cd2···Cd3 = 7.83 and Cd3···Cd1 = 6.77 Å.

The trinuclear cations [{Cd(tren)}$_3$(tren)]$^{6+}$ are organized in layers in the crystallographic ab plane and within one layer the arms of the tripodal cations are orientated in the same direction (Figure 2).

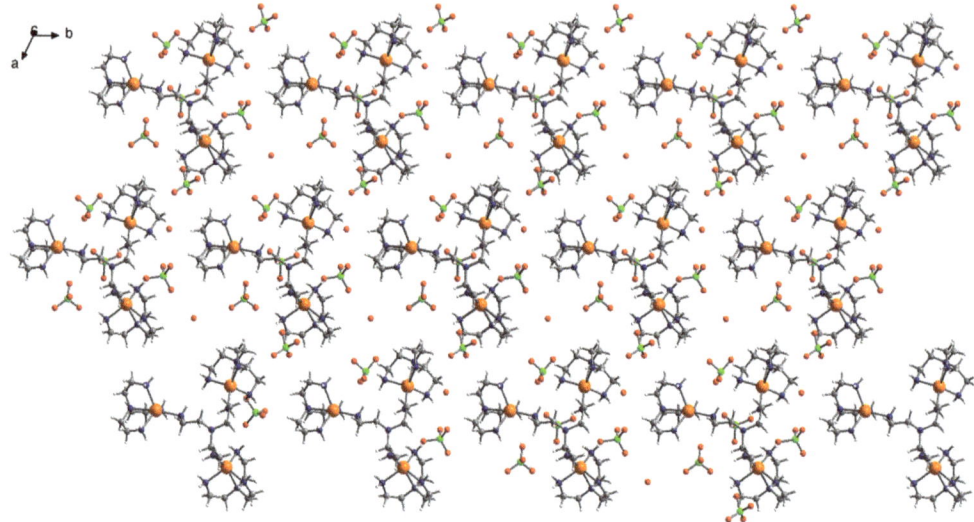

Figure 2. View along the crystallographic c axis of a packing diagram in crystal 1.

The exposure of the reaction mixture for longer time to atmospheric condition allows the partial replacement of the bridging tren ligands with carbonate anions resulted from the capture of atmospheric CO_2. Compound **2**, [{Cd(tren)}$_3$(tren)][{Cd(tren)}$_3$(μ_3-CO_3)](ClO$_4$)$_{10}$, crystallizes after more than one week and its structure consists in two types of trinuclear cations, [{Cd(tren)}$_3$(tren)]$^{6+}$ and [{Cd(tren)}$_3$(μ_3-CO_3)]$^{4+}$ (Figure 3), and perchlorate anions.

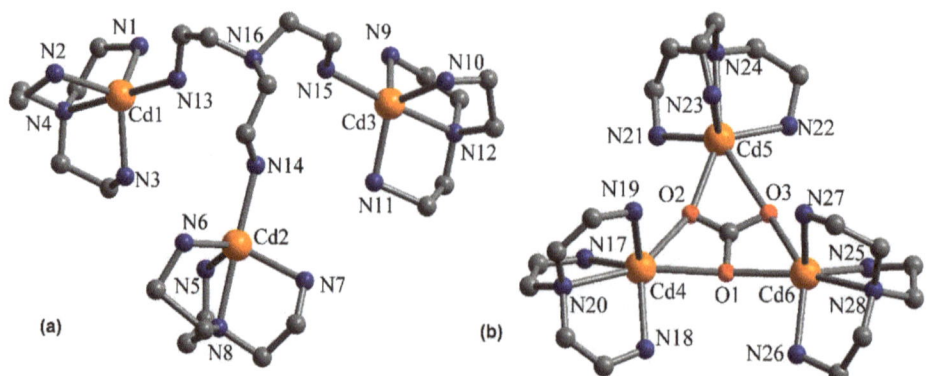

Figure 3. Perspective views of the two cationic complexes of the compound **2**, [{Cd(tren)}$_3$(tren)]$^{6+}$ (**a**) and [{Cd(tren)}$_3$(μ_3−CO_3)]$^{4+}$ (**b**). The hydrogen atoms were omitted for clarity.

The cation [{Cd(tren)}$_3$(tren)]$^{6+}$ preserves its general features from the compound **1** (Figure 3a). The Cd(II) ions from this cation are also pentacoordinated with a trigonal bipyramidal stereochemistry. The equatorial Cd-N bond lengths are: Cd1-N1 = 2.296(19), Cd1-N2 = 2.283(14), Cd1-N3 = 2.258(14), Cd2-N5 = 2.43(2), Cd2-N6 = 2.24(2), Cd2-N7 = 2.21(2), Cd3-N9 = 2.289(16), Cd3-N10 = 2.215(12) and Cd3-N11 = 2.236(15) Å, while the axial Cd-N bond lengths are: Cd1-N4 = 2.402(11), Cd1-N13 = 2.265(13), Cd2-N8 = 2.354(13), Cd2-N14 = 2.305(17), Cd3-N12 = 2.397(13) and Cd3-N15 = 2.299(12) Å. The Cd$\cdots$Cd distances within the trinuclear complex are: Cd1$\cdots$Cd2 = 6.68, Cd2$\cdots$Cd3 = 7.04 and Cd3$\cdots$Cd1 = 7.80 Å.

In the second trinuclear cationic complex, [{Cd(tren)}$_3$(μ_3-CO$_3$)]$^{4+}$, the carbonate anion coordinates in the μ_3-k^2:k^2:k^2 bridging tris-chelating fashion (Figure 3b). The coordination number for the three Cd(II) ions is six, the axial amino group of the bridging tren molecule being replaced by two chelating oxygen atoms of the carbonate anion. In the cation [{Cd(tren)}$_3$(μ_3-CO$_3$)]$^{4+}$ the Cd-N bond lengths are: Cd4-N17 = 2.278(14), Cd4-N18 = 2.307(14), Cd4-N19 = 2.276(11), Cd4-N20 = 2.471(9), Cd5-N21 = 2.292(13), Cd5-N22 = 2.260(12), Cd5-N23 = 2.261(12), Cd5-N24 = 2.453(10); Cd6-N25 = 2.290(12), Cd6-N26 = 2.279(11), Cd6-N27 = 2.281(10) and Cd6-N28 = 2.434(10) Å, while the Cd-O bond lengths are: Cd4-O1 = 2.429(8), Cd4-O2 = 2.430(8), Cd5-O2 = 2.327(8), Cd5-O3 = 2.629(9), Cd6-O1 = 2.347(8) and Cd6-O3 = 2.556(9) Å. The Cd$\cdots$Cd distances within the tetracationic trinuclear complex are shorter: Cd4$\cdots$Cd5 = 4.66, Cd5$\cdots$Cd6 = 5.11 and Cd6$\cdots$Cd4 = 4.68 Å.

The two types of trinuclear cations, [{Cd(tren)}$_3$(tren)]$^{6+}$ and [{Cd(tren)}$_3$(μ_3-CO$_3$)]$^{4+}$, are also arranged in layers, but in the crystallographic *bc* plane (Figure 4). In these layers, each type of trinuclear cation forms columns running along the crystallographic *b* axis with an alternating distribution of the two types of columns within the layer. Similarly to compound **1**, the arms of the tripodal cations [{Cd(tren)}$_3$(tren)]$^{6+}$ are orientated on the same direction within one layer and in opposite directions in the neighboring layers.

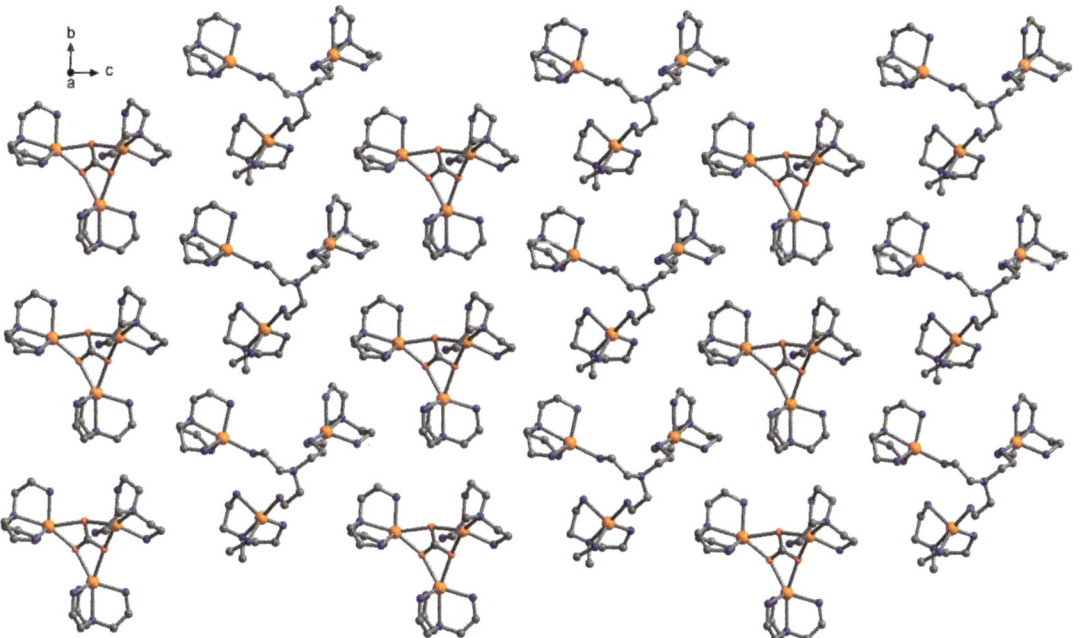

Figure 4. View along the crystallographic *a* axis of a packing diagram in crystal **2** showing the alternating distribution of the types of cations, [{Cd(tren)}$_3$(tren)]$^{6+}$ and [{Cd(tren)}$_3$(μ_3-CO$_3$)]$^{4+}$, in the crystallographic *bc* plane. The hydrogen atoms and perchlorate anions were omitted for clarity.

3.2. Spectral Properties

The absorption spectra of compounds **1** and **2** were acquired in the wavelength range 200–1000 nm on solid samples (using the diffuse reflectance technique) and both compounds present absorption maxima around 370 nm.

The room temperature photoluminescence of compounds **1** and **2** was explored using excitation wavelengths from the 350–400 nm range. The emission spectra displayed asymmetric bands with maxima at 470 nm for both compounds. The highest intensity of these bands was obtained when λ_{ex} = 350 nm. Compound **2** also presented a maximum at 450 nm and a "shoulder" at around 500 nm. In the emission spectrum of **1** there are two "shoulders" at 450 and 500 nm. The presence of the "shoulder" at 450 nm in the emission spectrum of **1** may suggest the presence of impurities of compound **2**. The compounds were separated mechanically and we cannot exclude the presence of impurities because both compounds are colourless. The corresponding excitation spectra (λ_{em} = 470 nm) reveals complex bands within the 250–450 nm domain with maxima at 410 nm for **1** and 400 nm for **2** (Figure 5).

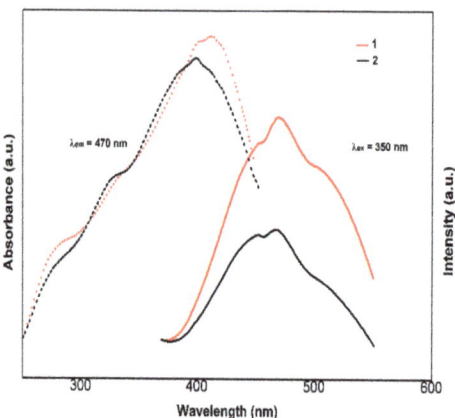

Figure 5. The solid-state emission (**right**) and excitation (**left**) spectra for the compounds **1** and **2**.

4. Conclusions

The cationic trinuclear Cd(II) complex with the tripodal tren ligand, $[\{Cd(tren)\}_3(tren)]^{6+}$, proved to be a good sequestering agent for atmospheric CO_2. The bridging tren ligand pre-organized the metal ions in a trigonal manner, favorable to a μ_3-k^2:k^2:k^2 coordination mode of the carbonate anion and this basic ligand is easy replaceable by the carbonate anion. The luminescent properties of the Cd(II) complexes may combine the sequestering abilities of such complexes with a sensing application.

Funding: This research was funded by UEFISCDI (Project PN-III-P2-2.1-PED-2019-2079, contract 469PED/2020).

Institutional Review Board Statement: Not applicable.

Informed Consent Statement: Not applicable.

Data Availability Statement: Not applicable.

Conflicts of Interest: The author declares no conflict of interest.

References

1. Palmer, D.A.; Van Eldik, R. The chemistry of Metal Carbonato and Carbon Dioxide Complexes. *Chem. Rev.* **1983**, *83*, 651–731. [CrossRef]
2. Sołtys-Brzostek, K.; Terlecki, M.; Sokołowski, K.; Lewiński, J. Chemical fixation and conversion of CO_2 into cyclic and cage-type metal carbonates. *Coord. Chem. Rev.* **2017**, *334*, 199–231. [CrossRef]

3. English, N.J.; El-Hendawy, M.M.; Mooney, D.A.; MacElroy, J.M.D. Perspectives on atmospheric CO_2 fixation in inorganic and biomimetic structures. *Coord. Chem. Rev.* **2014**, *269*, 85–95. [CrossRef]
4. Liu, B.; Jia, Y.-Y.; Jin, J.; Liu, X.-M.; Xue, G.-L. Layer structural bimetallic metamagnets obtained from the aggregation of $Ru_2(CO_3)_4^{3-}$ and Co^{2+} in existence of halogen. *Cryst. Eng. Comm.* **2013**, *15*, 4280–4287. [CrossRef]
5. Khan, S.; Roy, S.; Bhar, K.; Kumar, R.K.; Maji, T.K.; Ghosh, B.K. Syntheses, structures and properties of μ_3-carbonato bridged trinuclear zinc(II) complexes containing a tailored tetradentate amine. *Polyhedron* **2012**, *32*, 54–59. [CrossRef]
6. Ghionoiu, A.-E.; Popescu, D.-L.; Maxim, C.; Madalan, A.M.; Haiduc, I.; Andruh, M. Atmospheric CO_2 capture by a triphenyltin–1,2-bis(4-pyridyl)ethane system with formation of a rare trinuclear carbonato-centered core. *Inorg. Chem. Commun.* **2015**, *58*, 71–73. [CrossRef]
7. Peng, Y.-X.; Xu, F.; Yin, G.; Liu, Q.; Huang, W. Three trinuclear copper(II) complexes bridged by μ_3-CO_3^{2-} with different coordination modes. *J. Coord. Chem.* **2012**, *65*, 3949–3959. [CrossRef]
8. Kolks, G.; Lippard, S.J.; Waszczak, J.V. A Tricopper(II) Complex Containing a Triply Bridging Carbonate Group. *J. Am. Chem. Soc.* **1980**, *102*, 4832–4833. [CrossRef]
9. Guo, Y.-N.; Chen, X.-H.; Xue, S.; Tang, J. Molecular Assembly and Magnetic Dynamics of Two Novel Dy_6 and Dy_8 Aggregates. *Inorg. Chem.* **2012**, *51*, 4035–4042. [CrossRef]
10. Janzen, D.E.; Botros, M.E.; Van Derveer, D.G.; Grant, G.J. Fixation of atmospheric carbon dioxide by a cadmium(II) macrocyclic complex. *Dalton Trans.* **2007**, 5316–5321. [CrossRef]
11. Liang, X.; Parkinson, J.A.; Parsons, S.; Weishäupl, M.; Sadler, P.J. Cadmium Cyclam Complexes: Interconversion of *Cis* and *Trans* Configurations and Fixation of CO_2. *Inorg. Chem.* **2002**, *41*, 4539–4547. [CrossRef] [PubMed]
12. Bag, P.; Dutta, S.; Biswas, P.; Maji, S.K.; Flörke, U.; Nag, K. Fixation of carbon dioxide by macrocyclic lanthanide(III) complexes under neutral conditions producing self-assembled trimeric carbonato-bridged compounds with μ_3-η^2:η^2:η^2 bonding. *Dalton Trans.* **2012**, *41*, 3414–3423. [CrossRef]
13. Liu, H.-X.; Zhang, X.; Gao, X.-J.; Chen, C.; Huang, D. Synthesis and mechanism study of a dimeric tetranuclear carbonate-bridged copper(II) complex resulting from CO_2 fixation by controlling O_2 concentration. *Inorg. Chem. Commun.* **2016**, *68*, 63–67. [CrossRef]
14. Escuer, A.; El Fallah, M.S.; Kumar, S.B.; Mautner, F.; Vicente, R. Synthesis, crystal structure and magnetic behaviour of (μ_3-CO_3)[$Ni_3(Medpt)_3(NCSe)_4$], a new example of trinuclear nickel(II) complex with pentadentate carbonato bridge and strong antiferromagnetic coupling. *Polyhedron* **1998**, *18*, 377–381. [CrossRef]
15. Liu, C.-M.; Hao, X.; Zhang, D.-Q. CO_2-fixation into carbonate anions for the construction of 3d-4f cluster complexes with salen-type Schiff base ligands: From molecular magnetic refrigerants to luminescent single-molecule magnets. *Appl. Organomet. Chem.* **2020**, *5893*, 1–14. [CrossRef]
16. Escuer, A.; Vicente, R.; Kumar, S.J.; Mautner, F.A. Spin frustration in the butterfly-like tetrameric [$Ni_4(\mu$-$CO_3)_2(aetpy)_8$]-[ClO_4]$_4$ [aetpy = (2-aminoethyl)pyridine] complex. Structure and magnetic properties. *J. Chem. Soc. Dalton Trans.* **1998**, 3473–3477. [CrossRef]
17. Marvaud, V.; Decroix, C.; Scuiller, A.; Guyard-Duhayon, C.; Vaissermann, J.; Gonnet, F.; Verdaguer, M. Hexacyanometalate Molecular Chemistry: Heptanuclear Heterobimetallic Complexes; Control of the Ground Spin State. *Chem. Eur. J.* **2003**, *9*, 1677–1691. [CrossRef] [PubMed]
18. Luo, J.; Gao, Y.; Qiu, L.-J.; Liu, B.-S.; Zhang, X.-R.; Cui, L.-L.; Yang, F. Synthesis and characterization of nickel(II) and copper(II) tricyanomethanide complexes with tris(2-aminoethyl)amine as co-ligand. *Inorg. Chim. Acta* **2014**, *416*, 215–221. [CrossRef]
19. Perez-Toro, I.; Domínguez Martín, A.; Choquesillo-Lazarte, D.; Vílchez-Rodríguez, E.; Castiñeiras, A.; Niclós-Gutiérrez, J. Synthesis, thermogravimetric study and crystal structure of an N-rich copper(II) compound with tren ligands and nitrate counter-anions. *Thermochim. Acta* **2014**, *593*, 7–11. [CrossRef]
20. Klüfers, P.; Mayer, P. A Star-Shaped Heptanuclear Tetramine Cadmium(II) complex. *Acta Cryst.* **1998**, *54*, 722–725. [CrossRef]
21. Septavaux, J.; Tosi, C.; Jame, P.; Nervi, C.; Gobetto, R.; Leclaire, J. Simultaneous CO_2 capture and metal purification from waste streams using triple-level dynamic combinatorial chemistry. *Nat. Chem.* **2020**, *12*, 202–212. [CrossRef] [PubMed]

Article

Comparative Study of Cu/ZSM-5 Catalysts Synthesized by Two Ion-Exchange Methods

Dalia Santa Cruz-Navarro [1,*], Miguel Torres-Rodríguez [2], Mirella Gutiérrez-Arzaluz [2], Violeta Mugica-Álvarez [2] and Sibele Berenice Pergher [3]

[1] Posgrado en Ciencias e Ingeniería, Universidad Autónoma Metropolitana, Av. San Pablo 180, Azcapotzalco, Mexico City 02200, Mexico
[2] Área de Química Aplicada, Departamento de Ciencias Básicas, Universidad Autónoma Metropolitana, Av. San Pablo 180, Azcapotzalco, Mexico City 02200, Mexico; trm@azc.uam.mx (M.T.-R.); gam@azc.uam.mx (M.G.-A.); vma@azc.uam.mx (V.M.-Á.)
[3] Departamento de Química, Universidade Federal do Rio Grande do Norte, Av. Senador Salgado Filho, 3000 Centro de Convivência Djalma Marinho Sala 09 S/N Lagoa Nova, Natal 2408102, Brazil; sibelepergher@gmail.com
* Correspondence: sonysantacruz@gmail.com; Tel.: +52-55-2863-1693

Abstract: As catalysis is one of the pillars of green chemistry, this work aimed at continuing the development of synthesized catalysts under controlled conditions that allow the attainment of materials with the best physicochemical properties for the process for which they were designed. Based on this, the synthesis, characterization, and comparison of copper-based catalysts supported on ammonium and acidic ZSM-5-type zeolite by two ion exchange methods, liquid phase and solid state, are presented. The catalysts obtained were characterized by SEM/EDS, FTIR, XRD, and TPR to study the effect of the synthesis method on the physicochemical properties of each catalyst. The SEM/EDS results showed a homogeneous distribution of copper in the zeolite and the TPR led to determining the temperature ranges for the reduction of $Cu^{2+} \rightarrow Cu^+ \rightarrow Cu^0$. Furthermore, the X-ray results showed no modification of the structure of the zeolite after ion exchange, heat treatment, and TPR analysis.

Keywords: Cu-ZSM-5 catalysts; TPR; green chemistry

1. Introduction

Zeolites are microporous aluminosilicates that exhibit a framework composed by SiO_4 and AlO_4^- tetrahedral units [1]. Zeolites have been widely used as catalytic supports and catalysts due to their remarkable properties, such as molecular pore size, regular structure, and chemical and thermal stability, in addition to their ion exchange property, which gives them the ability to exchange their compensation cations for catalytically active metals, tuning their catalytic properties under controlled conditions [2,3]. Likewise, other factors to consider in obtaining zeolite-based catalysts are the preparation methods and the treatment conditions, which exert notable influence on the final properties of the catalysts [4]. Thus, a wide variety of methods have been reported, such as precipitation, impregnation, incipient wet impregnation, and ion exchange in the aqueous phase or in the solid state, among others [5–10]. The ion exchange method in the liquid phase and in the solid phase has shown advantages when exchanging the compensating cations of the zeolites for metal ions. This has led to the achievement of an excellent exchange of the compensating cation of the zeolite with the catalytically active metal ions, thus attaining an excellent amount of the metal within the pores, as well as a good metallic dispersion [10]. Solid-state ion exchange can create some active sites different from those obtained by the classical ion exchange method in the liquid phase [11], which can improve catalytic activity. In this regard, recently Gates et al. [12] and B. Gates [13] have published work on the importance of atomic dispersion (molecular single-site supported metal catalyst) of metals to improve catalytic activity. Additionally, Chen et al. [14] reported the importance of the metal ligand interaction of platinum atoms, coordinated with the oxygen atoms of MIL-101, which offers alternative routes to improving the selectivity in the hydrogenation of CO_2 to methanol. Additionally, theoretical and experimental studies on solid-phase ion exchange have been published; for example, Chen et al. [15], using calculations based on density functional theory, investigated the use of solid-state ion exchange of copper in zeolites, showing that the formation of a volatile metal complex facilitates the diffusion of the metal complex into the zeolite lattice, thus generating efficient ion exchange at a moderate temperature. The experimental study by Shwan et al. [9] shows the effect of different gaseous atmospheres during the solid-state ionic exchange of copper species in zeolites at 250 °C, obtaining active catalysts for the reduction of NO with NH_3, (NH_3-SCR). Additionally, Zhang et al. [16] have reported the use of solid-state ion exchange in the synthesis of dimethyl carbonate, using different types of zeolite: Cu-Y, Cu-ZSM-5, and Cu-MOR. Copper-based catalysts are interesting because of their low cost and their proven activity for oxidation-reduction reactions. Furthermore, the catalysts are mainly considered a pillar of green chemistry [17] and a basic element of sustainable processes.

Therefore, the objective of the present work is to carry out a comparative study of the physicochemical properties of Cu-ZSM-5 catalysts synthesized by two ion exchange methods, liquid phase (LPIE) and solid state (SSIE), SEM/EDS, FTIR, XRD, and TPR were used to verify if there were differences in the properties, to take them into account appropriately. The intention is to evaluate these catalysts in the transformation of CO_2 to products of added value forming a Cu-ZnO/Cu-zeolite-base hybrid catalyst and with the Cu-ZSM-5 catalysts to evaluate the oxidation of methane to methanol.

2. Materials and Methods

Commercial ammonium ZSM-5 zeolite, "Alfa Aesar, Ward Hill, MA, USA" SiO_2/Al_2O_3 at 30:1 molar ratio was used for the synthesis of the catalysts. The copper precursor salt was $Cu(NO_3)_2 \cdot H_2O$, 98%, "Sigma Aldrich, St. Louis, MA, USA". The ion exchange was carried out in both the solid-state and in the liquid phase, with the ammonium and acid zeolite forms.

2.1. Zeolite in Acid Form (H-ZSM-5)

To obtain the zeolite in acid form (H-ZSM-5), a sample of the ammonium zeolite ZSM-5 was calcined in a tube furnace (Thermolyne 21100 model, Dubuque, IA, USA) at 550 °C for 3 h, with a heating ramp of 5 °C/min under a nitrogen atmosphere.

2.2. Liquid Phase Ion Exchange (LPIE)

A sample of 1.0 g ZSM-5 zeolite was mixed with 15 mL of a 0.1 M solution of $Cu(NO_3)_2 \cdot H_2O$ in a rotary evaporator at 60 °C for 3 h at 160 rpm. Three ion exchange cycles were performed, with abundant deionized water washing followed by a drying step between each exchange at 80 °C for 12 h.

2.3. Solid-State Ion Exchange (SSIE)

The zeolite (1.0 g) was mixed with $Cu(NO_3)_2 \cdot H_2O$ salt (0.12 g) in an agate mortar. Subsequently, the mixture was placed in a quartz reactor and heated at 500 °C for 5 h, with a heating ramp of 5 °C/min in 5% NH_3/Ar, PRAXAIR at a constant flow of 15 mL/min.

2.4. Characterization

The catalysts were characterized by XRD, SEM/EDS, FTIR, and TPR.

The diffraction pattern of the catalysts was obtained through a Philips X-Ray Diffractometer (XRD), model X'Pert, in order to perform a qualitative analysis (phase identification). The diffractometer was fitted with a copper anode tube using K_α radiation and with an incident monochromator with λ = 1.5406 Å, a 2 theta range from 4.0° to 80°, a step size of 0.02°, and a step time of 2.5 s.

The morphological analysis of the catalysts was performed by Variable Pressure Scanning Electron Microscopy (SEM), Carl Zeiss instrument model Supra 55VP, and the elemental chemical analysis was performed by energy dispersive spectroscopy (SEM/EDS) using an Oxford detector.

The Fourier Transform Infrared Spectroscopy (FTIR) analysis was performed with a Varian spectrometer, Excalibur 3600 model, and the analysis was carried out by Attenuated Total Reflectance (ATR) with a range from 400 to 4000 cm^{-1} in the mid-infrared region.

The Temperature-Programmed Reduction (TPR) analysis was performed on a Bel Japan TPD/TPR/TPO analyzer, BelCat model, with two stages of pretreatment and reduction with hydrogen flow, using a quartz micro-reactor.

Pre-treatment: 70 mg of previously dried sample were weighed and heated at a rate of 10 °C/min until reaching 350 °C in an oxygen atmosphere (5% O_2/He, 50 cm^3 min^{-1}) for 30 min, and subsequently the sample was cooled in an Ar flow (50 cm^3 min^{-1}) to 50 °C, for 10 min with Ar flow.

Thermo reduction: With a heating ramp from 10 °C/min to 700 °C with a flow of reducing gas (5% H_2/Ar) of 40 cm^3 min^{-1}, the atmosphere and temperature were maintained for 30 min.

3. Results

Four catalysts were obtained as reported in Table 1.

Table 1. ZSM-5 zeolite supported copper catalysts.

Zeolite	Form	Ion Exchange	Catalyst
ZSM-5 molar ratio SiO_2/Al_2O_3 30:1	Ammonium	Liquid phase	ZA3L1
		Solid-state	ZA3S4
	Acid	Liquid phase	ZH3L1
		Solid-state	ZH3S3

3.1. XRD

In Figure 1a,b, the X-ray powder diffraction patterns display the typical lines of the ZSM-5 zeolite that can be clearly identified [18]. The zeolite acid solid-state (ZH3S3) catalyst shows the characteristic lines of metallic copper located at 2θ = 43.3, 50.4, and 74.1°, which correspond to planes (111), (200), and (220) [19,20]. This result suggests the presence of copper nanoparticles, due to the synthesis method which used a physical mixture of the precursor salt of copper with the zeolite, which is corroborated by the SEM results, of Section 3.1. In contrast to this result, the other catalysts do not present these diffraction lines, so copper is considered to be located within the pores and channels of the parent zeolite, out of the reach of the incident beam, after it was exchanged for the ammonium ion or, in the case of acidic zeolites, for the proton, forming a metal complex inside the channels [21]. Furthermore, it is important to highlight that the change in the intensity of the diffraction lines at low angles is more important for the samples synthesized by ionic exchange in solid state, compared to those from ionic exchange in the liquid phase. It is important to note that, the ion exchange method allows the exchange of the ammonium ion or the proton of the zeolite for the Cu^{2+} ions, leaving this ion tetrahedrally coordinated to the aluminum of the network that forms the zeolite channels [22].

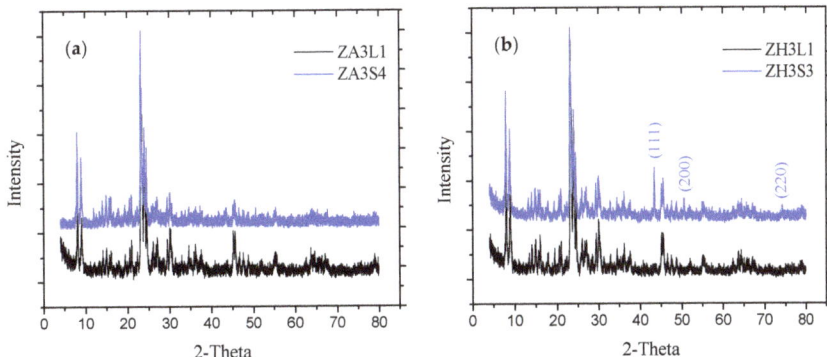

Figure 1. (**a**) Diffraction pattern of the catalysts ZA3L1 and ZA3S4; (**b**) Diffraction pattern of the catalysts ZH3L1 and ZH3S3.

3.2. SEM/EDS

Based on the results of the EDS analyses performed in triplicate (Figure 2), the catalysts obtained with the zeolite in acid form presented lower copper content than those of the ammonium zeolite. In addition, the catalysts obtained by SSIE displayed higher copper content than those obtained by LPIE.

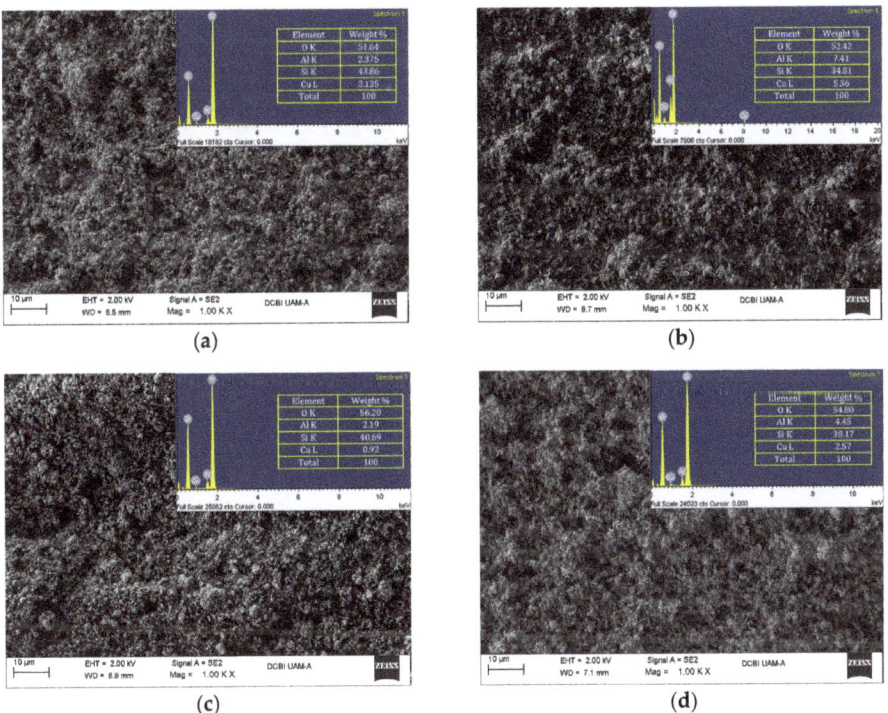

Figure 2. Micrographs and EDS analysis of the catalysts (**a**) ZA3L1, (**b**) ZA3S4, (**c**) ZH3L1, and (**d**) ZH3S3 that establish the overall granular morphology attained for each case.

The %w Cu in ZA3S4 (5.36%) > ZH3S3 (2.57%) > ZA3L1 (2.125%) > ZH3L1 (0.92%).

This difference in metal charge is attributed to the fact that SSIE facilitates the diffusion of the volatile metal complex into the zeolite pores when ion exchange is carried out at high temperature (500 °C) [9,11], which resulted in a more efficient exchange of the copper ion in the cationic centers located in the internal structure of the zeolite, and the achievement of a homogeneous metallic complex, as has recently been suggested [15].

In addition, in Figure 2 the SEM micrographs of the catalysts at a magnification of 1000× are presented, exhibiting nano-sized, homogeneous, irregularly shaped particles. Moreover, Figure 3 shows the elemental mapping of the ZA3S4 catalyst, in which the elements comprising it and their relative weight % are shown in tabular form in each inset. In the case of copper, the active catalyst phase shows a good distribution and dispersion of the metal in the zeolite, which indicates that the ion exchange method allowed a homogeneous metal distribution.

Figure 3. SEM secondary electron image of the ZA3S4 catalyst and four elemental mappings of the elements comprised.

3.3. FTIR

In Figure 4a, the IR spectra of the catalysts ZA3L1 and ZA3S4 are shown, and in Figure 4b, the IR spectra of the catalysts ZH3L1 and ZH3S3 are shown. In the spectra, only the characteristic absorption bands of the zeolites are observed, indicating the absence of elements or compounds foreign to the structure of the support after ion exchange with both methods, liquid phase and solid phase. The type of vibration assigned to each absorption band of the support is shown in Table 2.

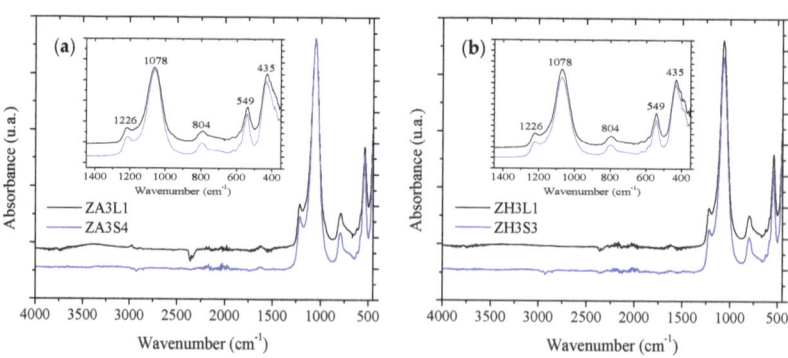

Figure 4. FTIR spectra of catalysts (**a**) ZA3L1 and ZA3S4, (**b**) ZH3L1 and ZH3S3, in the region of 4000 to 450 cm^{-1}.

Table 2. Type of vibrations identified in the FTIR spectrum of the catalysts [23,24].

Wave Number (cm^{-1})	Vibration Type
1226	External Asymmetric stretch (TO$_4$)
1078	Internal Asymmetric stretch (TO$_4$)
804	External Symmetrical stretch (TO$_4$)
549	External External link vibrations (TO$_4$)
435	Internal Bending of the tetrahedral (TO$_4$)
TO$_4$: Tetrahedral TO$_4$ (T = Si o Al)	

3.4. H_2-TPR

Figure 5a,b show the hydrogen reduction thermograms of the four catalysts, reported in Table 1. The reduction peaks for each catalyst were identified by the numbers in the thermograms. Three reduction peaks appear in the case of the catalysts obtained from zeolite in ammonium form (Figure 5a), whose presence has been reported for Cu/Zeolite ZSM-5 catalysts obtained by SSIE [25]. The reduction profiles of these two catalysts are similar to one another, except for the lag in reduction temperatures. For the catalysts obtained with zeolite in acid form, there were two reduction peaks (Figure 5b), which have also been reported and identified [26]; it is straightforward that the reduction profiles of these two catalysts are different from each other. The shape and position of the peaks vary and depend on the dispersion of Cu and the interaction with the zeolite matrix, as reported by Bulánek et al. [21].

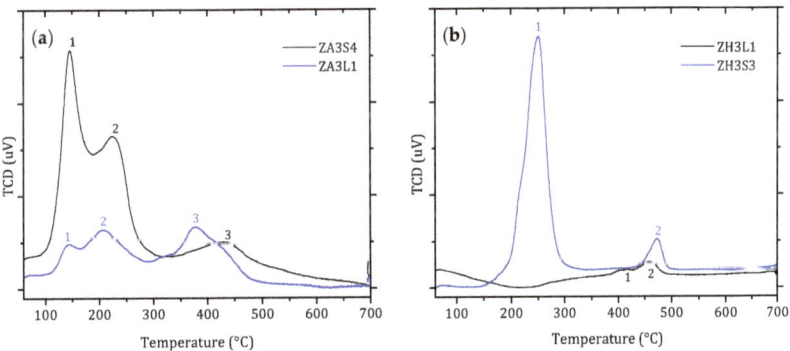

Figure 5. TPR profile (**a**) ZA3S4 and ZA3L1, (**b**) ZH3S3 and ZH3L1.

Table 3 shows the temperatures at which the maximum reduction peaks identified occurred for each catalyst. The moles of H$_2$ consumed per gram of catalyst are shown in Table 4, in which that corresponding to the catalyst ZA3S4 was the highest consumption of H$_2$ (6.403 mmol H$_2$/g). This is interpreted as the greater amount of copper present in this catalyst with respect to the others; it is observed that this result agrees with the elemental analysis results presented previously (% w Cu of 5.36%).

Table 3. Reduction temperatures of the Cu/zeolite ZSM-5 catalysts.

Peak No.	Reduction Temperatures (°C)			
	ZA3S4	ZA3L1	ZH3S3	ZH3L1
1	145.9	140.0	246.5	409.1
2	224.4	203.6	471.7	455.5
3	435.0	378.7	-	-

Table 4. H_2 consumed during TPR analysis on Cu/ZSM-5 catalysts.

Catalyst	mmol H_2	mmol H_2/g
ZA3S4	0.452	6.403
ZA3L1	0.217	3.098
ZH3S3	0.232	3.313
ZH3L1	0.014	0.206

The lower reduction temperatures of the catalysts obtained by LPIE with respect to the catalysts obtained by SSIE (Table 3) is attributed to a lower binding force of Cu^{2+} with the zeolite matrix; while, in the case of catalysts obtained by the SSIE method, the exchange is carried out mainly by gaseous diffusion of the metal through the pores, which allows the exchange of Cu within the less accessible sites and therefore requires higher temperatures for its reduction.

Regarding the copper species present, it is known that for low metal contents, the main species is Cu^{2+}, which can be reduced in two steps to Cu^0. The reduction of Cu^{2+} to Cu^+ occurs at low temperatures, while the reduction of Cu^+ to Cu^0 occurs at higher temperatures [27]. According to previous studies by Tounsi et al. and Delahay et al. [25,28], the reactions involved in the reduction process of copper-exchanged zeolites are the following:

$$CuO + H_2 \rightarrow Cu^0 + H_2 \tag{1}$$

$$Cu^{2+} + H_2 \rightarrow Cu^+ + 2H^+ \tag{2}$$

$$Cu^+ + H_2 \rightarrow Cu^0 + 2H^+ \tag{3}$$

Based on the above, in the case of the ZA3S4 and ZA3L1 catalysts, the Cu^{2+} species reduced to Cu^+ are associated with the first observed peak (reaction (2)), Cu^{2+} in CuO species reduced to Cu^0 with the second peak (reaction (1)) and species Cu^+ to Cu^0 with the third peak (reaction (3)) [25].

On the other hand, for the ZH3S3 catalyst, the peak of 246.5 °C is associated with the reduction of Cu^{2+} ions to Cu^+ (reaction (2)) and the second peak at 471.7 °C corresponds to the reduction of Cu^+ to Cu^0 (reaction (3)). Finally, for the ZH3L1 catalyst, the peaks greater than 350 °C correspond to the reduction of Cu^+ to Cu^0 [21,29].

4. Conclusions

The results show that the synthesis method of the catalysts established a difference in the resulting physical-chemical properties of each one. The EDS analyses showed that the catalysts obtained by SSIE exhibited a higher copper load than those obtained by LPIE, which is attributed to the fact that SSIE facilitates the diffusion of the volatile copper metal complex through the interior pores of the zeolite. The highest copper content was reached in the ZA3S4 catalyst (5.36% by weight), which was obtained by SSIE with the zeolite in ammonia form.

On the other hand, the XRD results show no difference between the catalysts and their reference standard, except for the ZH3S3 catalyst in which the characteristic diffraction lines of metallic copper are observed.

In the case of the FTIR analysis, the characteristic absorption bands of the ZSM-5 zeolite were observed for all the catalysts; however, it is not possible to visualize other absorption bands in the studied region that may be associated with the incorporation of the metal. Finally, by means of the TPR analysis, three peaks corresponding to the reduction of copper in the catalysts were obtained with ammonia zeolite, and two reduction peaks for the catalysts obtained with the zeolite in acid form were identified. The peaks correspond to the reduction of Cu^{2+} to Cu^+, CuO to Cu^0, and Cu^+ to Cu^0, in order of increasing temperature. The lower temperature reduction peaks were for the catalysts obtained by LPIE, rather than for those obtained by SSIE.

The results found suggest that the two ion exchange methods are suitable for incorporating Cu with a homogeneous distribution into the zeolite structure, with SSIE being the one that provides the highest exchange percentage. These catalysts can be used in the transformation of methane to methanol and the catalytic hydrogenation of CO_2 through a hybrid catalyst.

Author Contributions: D.S.C.-N. carried out the experimental work of synthesis of the Cu/ZSM-5 catalysts by two ion exchange methods, M.G.-A. tested and interpreted the SEM/EDS analysis of catalysts, V.M.-Á. conducted the discussion of results, M.T.-R. and S.B.P. participated in the direction and discussion of the results. All authors have read and agreed to the published version of the manuscript.

Funding: This research received no external funding.

Institutional Review Board Statement: Not applicable.

Informed Consent Statement: Not applicable.

Data Availability Statement: Data are contained within the article.

Acknowledgments: Gratefully acknowledge the financial support provided by Autonomous Metropolitan University and the National Council of Science and Technology (CONACyT) for the doctoral scholarship (CVU: 749385) and recognition to the SNI for the distinction of GAM, MAV and TRM membership and the stipend received.

Conflicts of Interest: The authors declare no conflict of interest.

References

1. Breck, D.W. *Zeolite Molecular Sieves: Structure, Chemistry and Use*; John Wiley & Sons: Hoboken, NJ, USA, 1974.
2. Yang, H.; Zhang, C.; Gao, P.; Wang, H.; Li, X.; Zhong, L.; Wei, W.; Sun, Y. A review of the catalytic hydrogenation of carbon dioxide into value-added hydrocarbons. *Catal. Sci. Technol.* **2017**, *7*, 4580–4598. [CrossRef]
3. Perego, C.; Carati, A. Zeolites and zeolite-like materials in industrial catalysis. In *Zeolites: From Model Materials to Industrial Catalysts*; Transworld Research Network: Kerala, India, 2008; pp. 357–389.
4. Liu, X.M.; Lu, G.; Yan, Z.F.; Beltramini, J. Recent Advances in Catalysts for Methanol Synthesis via ydrogenation of CO and CO_2. *Ind. Eng. Chem. Res.* **2003**, *42*, 6518–6530. [CrossRef]
5. Perego, C.; Pierluigi, V. Catalyst preparation methods. *Catal. Today* **1997**, *34*, 281–305. [CrossRef]
6. Pinna, F. Supported metal catalysts preparation. *Catal. Today* **1998**, *41*, 129–137. [CrossRef]
7. Tamas, I.; Ganea, R.; Pop, G. Catalysts and reaction conditions screening by microreactor devices for exhaust gas purification. In *Microreaction Technology*; Springer: Berlin/Heidelberg, Germany, 2001; pp. 464–469.
8. Campanati, M.; Fornasari, G.; Vaccari, A. Fundamentals in the preparation of heterogeneous catalysts. *Catal. Today* **2003**, *77*, 299–314. [CrossRef]
9. Shwan, S.; Skoglundh, M.; Lundegaard, L.; Tiruvalam, R.; Janssens, T.; Carlsson, A.; Vennestrøm, P. Solid-state ion-exchange of copper into zeolites facilitated by ammonia at low temperature. *ACS Catal.* **2015**, *5*, 16–19. [CrossRef]
10. Jouini, H.; Mejri, I.; Petitto, C.; Martínez, J.; Vidal, A.; Mhamdi, M.; Blasco, T.; Delahay, G. Characterization and NH_3-SCR reactivity of Cu-Fe-ZSM-5 catalysts prepared by solid state ion exchange: The metal exchange order effect. *Microporous Mesoporous Mater.* **2018**, *260*, 217–226. [CrossRef]
11. Abu, B.; Schwieger, W.; Unger, A. Nitrous oxide decomposition over transition metal exchanged ZSM-5 zeolites prepared by the solid-state ion exchange method. *Appl. Catal. B Environ.* **2008**, *84*, 277–288.
12. Gates, B.C.; Flytzani-Stephanopoulos, M.; Dixon, D.A.; Katz, A. Atomically dispersed supported metal catalysts: Perspectives and suggestions for future research. *Catal. Sci. Technol.* **2017**, *7*, 4259–4275. [CrossRef]
13. Gates, B.C. Atomically Dispersed Supported Metal Catalysts: Seeing Is Believing. *Trends Chem.* **2019**, *1*, 99–110. [CrossRef]
14. Chen, Y.; Li, H.; Zhao, W.; Zhang, W.; Li, J.; Li, W.; Zheng, X.; Yan, W.; Zhang, W.; Zhu, J.; et al. Optimizing reaction paths for methanol synthesis from CO_2 hydrogenation via metal-ligand cooperativity. *Nat. Commun.* **2019**, *10*, 1885. [CrossRef] [PubMed]

15. Chen, L.; Jansson, J.; Skoglundh, M.; Grönbeck, H. Mechanism for Solid-State Ion Exchange of Cu+ into Zeolites. *J. Phys. Chem. C* **2016**, *120*, 29182–29189. [CrossRef]
16. Zhang, Y.; Briggs, D.N.; de Smit, E.; Bell, A.T. Effects of zeolite structure and composition on the synthesis of dimethyl carbonate by oxidative carbonylation of methanol on Cu-exchanged Y, ZSM-5, and Mordenite. *J. Catal.* **2007**, *251*, 443–452. [CrossRef]
17. Anastas, P.; Kirchhoff, M.; Williamson, T. Catalysis as a foundational pillar of green chemistry. *Appl. Catal. A Gen.* **2001**, *221*, 3–13. [CrossRef]
18. International Zeolite Association. Database of Zeolite Structures. Structure Commission of the International Zeolite Association. 2017. Available online: http://america.iza-structure.org/IZA-SC/pow_pat.php?STC=MFI&ID=MFI_0 (accessed on 10 September 2021).
19. Natesakhawat, S.; Lekse, J.; Baltrus, J.; Ohodnicki, P.; Howard, B.; Deng, X.; Matranga, C. Active sites and structure−activity relationships of copper-based catalysts for carbon dioxide hydrogenation to methanol. *ACS Catal.* **2012**, *2*, 1667–1676. [CrossRef]
20. Li, S.; Guoa, L.; Ishiharab, T. Hydrogenation of CO_2 to methanol over Cu/AlCeO catalyst. *Catal. Today* **2020**, *339*, 352–361. [CrossRef]
21. Bulánek, R.; Wichterlová, B.; Sobalík, Z.; Tichý, J. Reducibility and oxidation activity of Cu ions in zeolites Effect of Cu ion coordination and zeolite framework composition. *Appl. Catal. B Environ.* **2001**, *31*, 13–25. [CrossRef]
22. Drake, I.J.; Zhang, Y.; Briggs, D.; Lim, B.; Chau, T.; Bell, A. The Local Environment of Cu^+ in Cu-Y Zeolite and Its Relationship to the Synthesis of Dimethyl Carbonate. *J. Phys. Chem. B* **2006**, *110*, 11654–11664. [CrossRef]
23. Li, C.; Wu, Z.; Auerbac, S.; Carrado, K.; Dutta, P. Microporous materials characterized by vibrational spectroscopies. In *Handbook of Zeolite Science and Technology*; CRC Press: Boca Raton, FL, USA, 2003; pp. 435–436.
24. Kulprathipanja, S. Zeolite Characterization. In *Zeolites in Industrial Separation and Catalysis*; John Wiley & Sons: Hoboken, NJ, USA, 2010; p. 114.
25. Tounsi, H.; Djemel, S.; Ghorbel, A.; Delahay, G. Characterization and performance of over-exchanged Cu-ZSM-5 catalysts prepared by solid-state ion exchange for the selective catalytic reduction of no by n-decane. *J. Société Chim. Tunis.* **2007**, *9*, 85–96.
26. Nanba, T.; Masukawa, S.; Ogata, A.; Uchisawa, J.; Obuchi, A. Active sites of Cu-ZSM-5 for the decomposition of acrylonitrile. *Appl. Catal. B Environ.* **2005**, *61*, 288–296. [CrossRef]
27. De Lucas, A.; Valverde, J.L.; Dorado, F.; Romero, A.; Asencio, I. Influence of the ion exchanged metal (Cu, Co, Ni and Mn) on the selective catalytic reduction of NOX over mordenite and ZSM-5. *J. Mol. Catal. A Chem.* **2004**, *225*, 47–58. [CrossRef]
28. Delahay, G.; Coq, B.; Broussous, L. Selective catalytic reduction of nitrogen monoxide by decane on copper-exchanged beta zeolites. *Appl. Catal. B Environ.* **1996**, *12*, 49–59. [CrossRef]
29. Urquieta González, E.; Martins, L.; Peguin, R.P.S.; Batista, M.S. Identification of Extra-Framework Species on Fe/ZSM-5 and Cu/ZSM-5 Catalysts Typical Microporous Molecular Sieves with Zeolitic Structure. *Mater. Res.* **2002**, *5*, 321–327. [CrossRef]

Article

Performance Enhancement of Self-Cleaning Cotton Fabric with ZnO NPs and Dicarboxylic Acids

Xinlei Ji, Hong Li *, Yuan Qin and Jun Yan

Department of Textile and Material Engineering, Dalian Polytechnic University, Dalian 116034, China; jxl19980119@163.com (X.J.); qinyd163@163.com (Y.Q.); yanjun@dlpu.edu.cn (J.Y.)
* Correspondence: lihong@dlpu.edu.cn

Abstract: In this paper, we explore the self-cleaning and washing durability of green-prepared ZnO NPs combined with cotton fabrics. Honeysuckle extract was used to prepare ZnO NPs with an average particle size of 15.3 nm. Cotton fabrics were then treated with oxalic acid (OA), tartaric acid (TA), and succinic acid (SA) as cross-linking agents, sodium hypophosphite as a catalyst, and after that, the ZnO NPs were applied to the cross-linked cotton fabrics by the padding to prepare the self-cleaning cotton fabrics. The morphology and structure of the fabric samples were characterized using FTIR, scanning electron microscopy (SEM), Energy-dispersive X-ray spectroscopy (EDS), and XRD. The optical properties of the cotton fabric samples were discussed by UV-vis diffuse reflectance spectrum, and the self-cleaning performance, wrinkle recovery angle and ultraviolet protection performance of the cotton fabric samples were analyzed. The results showed that the carboxyl groups of TA, OA, and SA were esterified with hydroxyl groups of the cotton fiber and formed a film on the surface of the cotton fabrics. ZnO NPs were successfully loaded onto the cotton fabrics by strong electrostatic interaction, causing the improvement of the washing resistance of the cross-linked fabrics. In addition, compared with uncross-linked fabrics, the wrinkle recovery performance of the cross-linked fabrics had also been greatly improved, and the UV protection factor reached 50+, thus obtaining an excellent self-cleaning, multifunctional cotton-based textile with anti-wrinkle and anti-ultraviolet properties.

Keywords: cotton fabrics; ZnO NPs; cross-linked; washing resistance; self-cleaning

Citation: Ji, X.; Li, H.; Qin, Y.; Yan, J. Performance Enhancement of Self-Cleaning Cotton Fabric with ZnO NPs and Dicarboxylic Acids. *Crystals* **2022**, *12*, 214. https://doi.org/10.3390/cryst12020214

Academic Editor: Shujun Zhang

Received: 27 December 2021
Accepted: 29 January 2022
Published: 31 January 2022

Publisher's Note: MDPI stays neutral with regard to jurisdictional claims in published maps and institutional affiliations.

Copyright: © 2022 by the authors. Licensee MDPI, Basel, Switzerland. This article is an open access article distributed under the terms and conditions of the Creative Commons Attribution (CC BY) license (https://creativecommons.org/licenses/by/4.0/).

1. Introduction

Cotton fabrics are widely used in clothing and household products because of their good moisture absorption, breathability, and skin affinity [1]. However, cotton fabrics also have shortcomings such as easy staining, easy wrinkling, and poor UV resistance [2]. Bacteria easily breed on the surface of stained fabrics, leading to an increased risk of illness for users. The formation of wrinkles affects aesthetics and fabric comfort. Cotton fabrics have poor UV resistance, which accelerates the skin aging of users and even induces skin cancer. Therefore, it is particularly necessary to endow cotton fabrics with a certain degree of self-cleaning, anti-wrinkle, and anti-ultraviolet properties.

Nano ZnO is an inorganic metal oxide, which has non-toxicity and low cost. There are many ways to synthesize ZnO, including the hydrothermal method, sol-gel method, and the precipitation method [3–6]. However, most of these methods involve using toxic chemicals and solvents harmful to human health and the environment. At present, the green preparation of nanoparticles with plant extracts such as aloe extract [7] and viburnum extract [8] have become a simple and eco-friendly preparation method. It was reported that green-prepared ZnO NPs have a smaller particle size and better optical properties [9].

Nano ZnO is also a semiconductor with a wide bandgap and high excitation binding energy [10]. When the nanoparticle is irradiated by light with energy greater than or equal to its forbidden bandwidth, the electrons are excited and transferred from the valence band to the conduction band, forming conduction band electrons (e^-), while the valence band leaves holes (h^+), then photo-generated electrons and holes are respectively captured by O_2 and H_2O molecules adsorbed on the surface, and finally, generate hydroxyl radicals ($\cdot OH$). The free radicals have an oxidation potential of up to 2.7 eV, which has strong oxidizing properties and can indiscriminately oxidize and decompose the adsorbed pollutants [3,11]. So the incorporation of ZnO NPs with cotton fabric or polyester has been reported to develop photocatalytic self-cleaning textiles [12,13].

The washing resistance of functional textiles developed with nanoparticles is also a problem worthy of study. The use of cross-linking agents on the fabrics not only increases the functionality of fabrics but also improves the washing durability of fabrics. Dimethylol dihydroxy ethylene urea (DMDHEU) is one of the best cross-linkers for cotton, but it releases a high amount of toxic formaldehyde [14]. The polycarboxylic acid cross-linking agent is a formaldehyde-free crosslinking agent, which can not only give cotton fabrics easy-care performance [15,16] but also effectively connect nanoparticles to cotton fabrics [17]. So, it is the most commonly used cross-linking agent. Majid has illustrated butane tetracarboxylic acid (BTCA) as a cross-linking agent to enhance the stability of Ag NPs on cotton fabrics [18]. However, the high cost and not cost-efficient of BTCA have lost its commercial viability; J L Zhu prepared the anti-wrinkle and antibacterial cotton fabric with good washing resistance by one-liquor finishing using a combination of polymaleic acid (PMA) and citric acid (CA) and Ag NPs [19]; Loghman Karimi adopted SA as a cross-linking agent to attach TiO_2 on cotton, and revealed that the self-cleaning degree of the samples treated by the cross-linking was much higher than that of the non-cross-linked cotton [20].

Some studies have been done on the application of ZnO NPs in the development of self-cleaning fabrics [21], but there are few studies on the use of different dicarboxylic acids as cross-linking agents to immobilize ZnO NPs to improve their washing resistance. So, in this paper, a variety of dicarboxylic acids were used as cross-linking agents to immobilize ZnO NPs, and they were compared in detail to find the best dicarboxylic acid. First, ZnO NPs were prepared by plant extract-honeysuckle extract. Then three dicarboxylic acids (tartaric acid [TA], oxalic acid [OA], and succinic acid [SA]) were used to finish cotton fabrics using sodium hypophosphite (SHP) as the catalyst. Finally, the cotton fabrics were padded with a ZnO NPs solution. In this way, the cotton fabric not only obtained long-term self-cleaning properties but also greatly improved its anti-wrinkle and anti-ultraviolet properties.

2. Experimental

2.1. Materials

Pure cotton bleached woven fabrics (440 picks/10 cm × 421 picks/10 cm) was obtained by Qingdao Coastal Textile Co., LTD., Shandong, China. Honeysuckle extract (80 mesh powder, containing 50% chlorogenic acid) was obtained by Shandong Deyan Chemical Co., LTD, Shandong, China. Zinc Nitrate Hexahydrate ($ZnNO_3 \cdot 6H_2O$) was obtained by Shanghai Guangrui Biotechnology Co., LTD., Shanghai, China. Methylene blue, Oxalic acid (OA), Tartaric acid (TA), succinic acid (SA), sodium hypophosphite monohydrate (SHP, $NaH_2PO_2 \cdot H_2O$), as well as other chemicals were of reagent grade was obtained by Tianjin Cormier Chemical Reagent Co., LTD., Tianjin, China.

2.2. Preparation of Self-Cleaning Cotton Fabric

2.2.1. Preparation of ZnO NPs

Zinc Nitrate Hexahydrate was used as the source of zinc in the experiment. First, Honeysuckle extract (1 g) was dissolved in 50 mL deionized water, centrifuged at 6000 r/min for 10 min, and the supernatant was collected for use as honeysuckle extract solution. Secondly, 3 mL honeysuckle extract solution and 10 mL Zinc Nitrate Hexahydrate solution of 0.1 mol/L were added into 37 mL deionized water and adjusted the pH to 7 with 0.1 mol/L NaOH solution, then heated in an oil liquor at 140 °C for 60 min to obtain a light yellow complex. Finally, the complexes were calcined at 300 °C for 3 h in a muffle furnace (atmosphere condition was air) to obtain ZnO NPs powders. Under the same conditions, a control experiment to prepare ZnO NPs without the addition of honeysuckle extract was carried out. Particle size analysis was performed by Transmission electron microscope (TEM).

2.2.2. Cross-Linking of Cotton Fabric with Dicarboxylic Acids

First, the cotton was pretreated (scoured with boiled water containing 2 g/L soap flakes and 2 g/L sodium carbonate at liquor ratio 1:30 [liquid ratio is a term used in textile dyeing and finishing, the ratio of the mass of cotton fabric to the volume of the solution, g: mL, same as below] for 5 min), and dried for later use. Secondly, the pretreated cotton fabrics were immersed in the prepared mixed solution of dicarboxylic acid (6% o.w.s.) and SHP (4% o.w.s.). The mixed solution was a mixed solution of binary carboxylic acid and catalyst sodium hypophosphate. The solvent is deionized water, and the solutes are dicarboxylic acid and SHP. The solvent is deionized water, and the solutes are dicarboxylic acid and SHP (liquor ratio 1:30—the mass ratio of cotton fabric to the solution), which is stirred with an oscillating frequency of 120 r/min at 50 °C for 60 min, then double-dipped and double-nipped [rolling residual rate (80 ± 1%), the same below]. The fabrics were dried at 70 °C for 30 min and cured at 160 °C for 2 min. Finally, the samples were washed with water for 5 min to remove excess reactants.

2.2.3. Finishing of Cotton Fabric with ZnO NPs

The cross-linked cotton fabrics with different dicarboxylic acids and a piece of uncross-linked cotton fabric were immersed into 0.005 mol/L ZnO NPs solution with a liquor ratio of 1:30, respectively, and stirred at 50 °C for 60 min. After two dips and two nips, the samples were put in an oven at 70 °C for 60 min and taken out for later use. The samples were labeled TA-ZnO/cotton, OA-ZnO/cotton, SA-ZnO/cotton, and ZnO/cotton.

2.3. Characterization of ZnO NPs and Fabrics
2.3.1. Structure Characterization

The X-ray diffraction patterns (XRD) of ZnO NPs and the cotton fabric samples were tested using an X-ray diffractometer. The test conditions: tube voltage 40 kV, tube current 30 mA, scanning speed: 5 (°)/min, 2θ: 10~90°, The crystallite size of ZnO was calculated by Scherrer's formula [12].

$$D = \frac{K\lambda}{\beta \cos \theta} \quad (1)$$

where K is the Scherrer constant, 0.89; λ is the X-ray wavelength at 0.1541 nm; D is the size of the ZnO NPs, and β is the width at half-maximum; θ is the Bragg diffraction angle (in degrees).

The morphology and particle size of Zn NPs were tested by Transmission electron microscope (JEM-2100 (UHR), Beijing, China), and the accelerating voltage was 200 kV.

Fourier-transform infrared (FTIR) spectra of fabric samples were recorded through an FTIR spectrometer (FTIR920, Suzhou, China), test condition: KBr tablet, wave number range: 4000~400 cm^{-1}. The surface morphologies of fabric samples were observed using a scanning electron microscope (SEM, Alpha300S, Dongguan, China);

The element of fabric samples was tested by an electronic energy spectrometer (EDS, HD-2001, Dongguan, China);

The content of the Zn element in the solution (wavelength band is 213 nm) was detected by an inductively coupled plasma atomic emission spectrometer (ICP-AES, Guangzhou, China), the experiment was repeated 5 times, and the average value was obtained.

The water contact angle of the fabrics was measured by an automatic surface tensiometer (k100c, Shanghai, China).

2.3.2. Optical Performance

Ultraviolet-visible (UV-vis) diffuse reflectance spectra of ZnO NPs and fabric samples were performed by a UV-vis spectrometer (Shimadzu, Japan), wave number range: 200~600 nm.

2.3.3. Self-Cleaning of Fabric

To estimate the self-cleaning of the cotton fabrics finished with ZnO NPs, methylene blue, coffee, red wine, and soy sauce were selected as target organic pollutants. 1 mL methylene blue (10 mg/L) solution, coffee, red wine, and soy sauce (commercially available) were dropped on the surface of 5 cm × 5 cm fabrics that were then placed in a dark place for 30 min. After the stable absorption, the K/S values of the fabrics were measured. Then they were exposed to simulated sunlight from an argon lamp that was 15 cm away from the fabrics for a period of time before the K/S values of the fabrics were measured once again. The catalytic degradation rates of the fabrics were calculated according to the K/S values of the fabrics before and after the light, as shown in Formula (2).

$$\zeta = \frac{(K/S)_0 - (K/S)_t}{(K/S)_0} \times 100\% \tag{2}$$

where ζ is the degradation rate; $(K/S)_0$ is the K/S value of the fabric before light; $(K/S)_t$ is the K/S after t hours of light.

To evaluate the self-cleaning durability of the fabrics, the fabrics stained by methylene blue were degraded with sunlight for 4 h, and the degradation rate was calculated. Then the fabrics were washed according to the AATCC-61-2006 method. The degradation rate was calculated after 1–10 times washing circles. The SEM images and content of the Zn element of the samples after 10 times of washing were carried out.

2.3.4. UV and Wrinkle Resistance of Fabric

The UV resistance of the ZnO/Cotton, OA-ZnO/cotton, TA-ZnO/cotton, and SA-ZnO/Cotton was measured using a Textile UV Performance Tester (YG(B)912E, Chengdu, China). The wrinkle recovery angle (WRA) of fabrics was tested using an automatic digital fabric crease elastomer (YG(B)541D-II, Xian, China).

3. Results and Discussion

3.1. Characterization and Analysis of ZnO NPs

The TEM of ZnO NPs prepared without the addition of honeysuckle extract is shown in Figure 1b. It can be seen that ZnO NPs have large-area aggregation, large particle size, and poor effect, so they cannot be used for subsequent experiments. By comparison, Figure 1a shows that the ZnO NPS prepared by adding honeysuckle extract is uniformly dispersed, with a smaller particle size and better effect [19].

X-ray powder diffractograms for ZnO NPs are shown in Figure 2. It can be seen that the positions of the diffraction peaks of 31.78°, 34.43°, 36.26°, 47.55°, 56.61°, 62.87°, 66.40°, 67.97°, 69.11°, 72.58°, 76.98°, and 81.40° correspond to crystal planes (100), (002), (101), (102), (110), (103), (200), (112), (201), (004), (202) and (104), respectively, which were consistent with the values in the standard card (JCPD 89-0510). The (101) plane exhibited the highest relative intensity for the entire XRD pattern, suggesting anisotropic growth and preferred orientation of the crystallites, a typical feature of wurtzite-structured materials [22,23]. In addition, no other peaks were observed in the XRD pattern, indicating the high purity of the powders. The average crystallite size was determined by Scherrer's equation 1 (Equation (1)), and the calculated crystallite size was 15.3 nm. The morphology of the ZnO NPs was characterized using TEM. It can be seen from Figure 1a (right) that most of the ZnO NPs prepared using the honeysuckle extract were approximately spherical, and the particle size from the TEM was smaller and close to the particle size calculated by the XRD pattern.

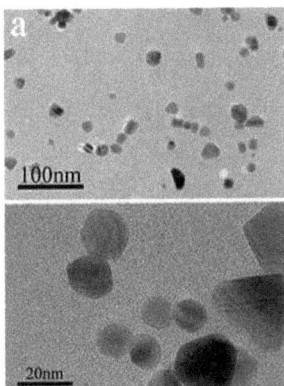
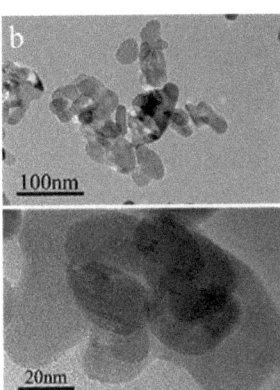

Figure 1. TEM images of ZnO NPs with honeysuckle extract added (**a**). TEM images of ZnO NPs without honeysuckle extract added (**b**).

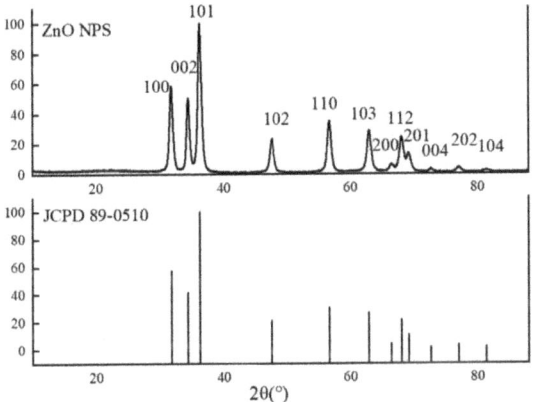

Figure 2. XRD patterns of ZnO NPs.

3.2. FTIR Analysis of Fabrics

FTIR analysis of the raw cotton, TA-ZnO/cotton, OA-ZnO/cotton, and SA-ZnO/cotton was performed in Figure 3. Cellulose is an important component of cotton fabrics, the macromolecular chain of cellulose mainly contains -OH, C-H, -CH_2-, and C-O functional groups. As shown in Figure 3a, the peak at 3410 cm^{-1} is attributed to the -OH stretching vibration peak; the peak at 2901 cm^{-1} is attributable to the C-H stretching vibration absorption peak; the peak at 1371 cm^{-1} belongs to the -CH_2- flexural vibration absorption peak; the peak at 1058 cm^{-1} is attributable to the C-O stretching vibration absorption peak; the peak at 1640 cm^{-1} belongs to the bending vibration peak of water adsorption [24]. It can be seen from the image of ZnO/cotton, TA-ZnO/cotton, OA-ZnO/cotton, and SA-ZnO/cotton in Figure 3b–e compared with the raw cotton that the peak at 435 cm^{-1} was stronger because the stretching vibration peak of Zn-O bond is between 420~450 cm^{-1}, showing that the cotton fabrics were loaded with ZnO NPs [25]. It can be seen from the images of TA-ZnO/cotton, OA-ZnO/cotton, and SA-ZnO/cotton in Figure 3c–e, a new peak appears at 1732 cm^{-1}, it is attributable to the peaks of ester and carboxyl groups. It is evident that hydroxyl of cotton fiber was esterified with carboxyl in dicarboxylic acids, and carboxyl was introduced [26].

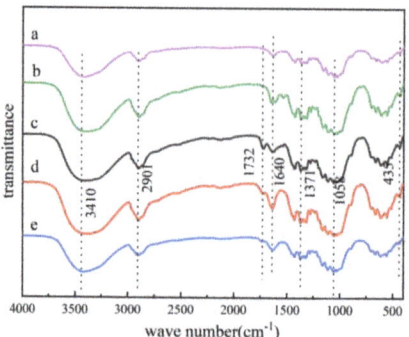

Figure 3. FTIR images of rawcotton (**a**), ZnO/cotton (**b**), OA-ZnO/cotton (**c**), TA-ZnO/cotton (**d**) and SA-ZnO/cotton (**e**).

3.3. SEM, EDS, and ICP-AES Analysis of Fabrics

The scanning electron microscopy (SEM) images of raw cotton (a), ZnO/cotton (b), OA-ZnO/cotton (c), TA-ZnO/cotton (d), and SA-ZnO/cotton (e) were presented in Figure 4. The content of the Zn element of cotton fabric loaded with ZnO NPs was measured by ICP-AES, and the result is shown in the upper right corner of the SEM image. The surface of the raw cotton fabric has many natural grooves, and it was not loaded with any particles. The surface of the ZnO/cotton fabric was loaded with a lot of spherical particles, and there was a slight agglomeration phenomenon. A layer of membrane structure formed on the OA-ZnO/cotton, TA-ZnO/cotton, and SA-ZnO/cotton fabrics, of which only a few spherical particles were on their surfaces, were evenly distributed and smaller in size.

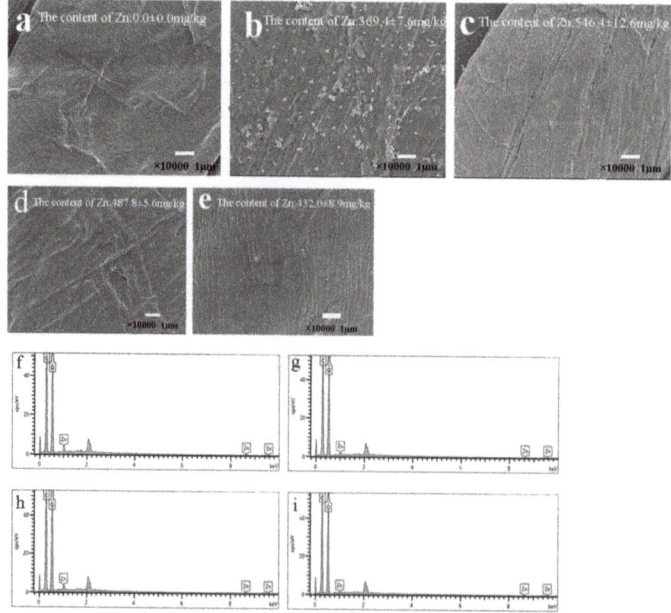

Figure 4. SEM images and Zn element contents of raw Cotton (**a**), ZnO/cotton (**b**), OA-ZnO/cotton (**c**), TA-ZnO/cotton (**d**), and SA-ZnO/cotton (**e**) and EDS spectra of ZnO/cotton (**f**), OA-ZnO/cotton (**g**), TA-ZnO/cotton (**h**) and SA-ZnO/cotton (**i**).

As shown in Figure 4f–i of Energy Dispersive Spectroscopy (EDS), zinc and oxygen were two elements on the finished samples apart from the carbon related to the cotton fabrics, verifying that the ZnO NPs were successfully loaded on the fabrics.

SEM revealed that the number of nanoparticles on the cross-linked fabric surfaces was less than on ZnO/cotton. However, the ICP-AEM data showed that the content of the Zn element in the cross-linked cotton fabrics was more than that in the ZnO/cotton. This indicates that the carboxyl groups of the cross-linked fabric adsorbed more ZnO NPs through strong electrostatic force; due to the relatively high electronegativity of the oxygen atom in the Zn-O bond of nano-zinc oxide, the electrons of zinc will be biased towards oxygen, which makes zinc positive. Then, because the oxygen on the carboxyl group shows a negative charge, the ZnO NPs are firmly loaded in the cotton fabric due to the electrostatic adsorption between anions and cations [27], and the ZnO NPs loaded on the cross-linked fabrics were mainly distributed under the film. In addition, the content of ZnO NPs on OA-ZnO/cotton (546.4 ± 12.6 mg/kg) was more than that on TA-ZnO/cotton (487.8 ± 5.6 mg/kg) and SA-ZnO/cotton (432.0 ± 8.9 mg/kg), which can be attributed to the more acidic potential and least pKa of oxalic acid (1.27) compared to that of two other dicarboxylic acids (pKa succinic acid = 4.21, pKa Tartaric acid = 3.04) [27]. The higher reactivity of oxalic acid resulted in a higher degree of esterification for the fabric, and as a result, the carboxyl content of fabric tended to increase, and the trend for electrostatic absorption of ZnO NPs increased [28]. The formation of ester cross-linkage of OA, TA, and SA with cellulose chain and ZnO and OA, TA, and SA linkage was shown in Figure 5 [29].

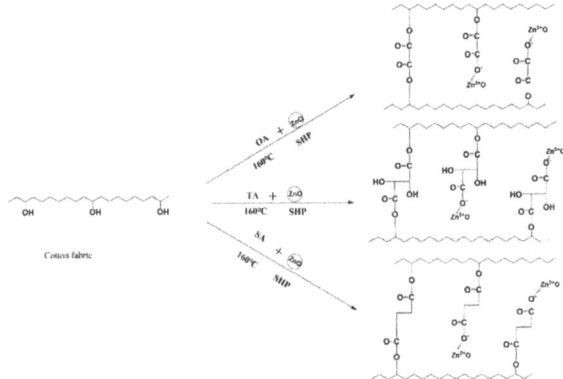

Figure 5. Schematic diagram of the cross-linked structure of OA, TA, and SA with cellulose chain and linkage of ZnO and OA, TA, and SA.

3.4. XRD Analysis of Fabrics

X-ray powder diffractograms of cotton fabrics before and after finishing with ZnO NPs are shown in Figure 6. It can be seen that the positions of the diffraction peaks of 14.95°, 16.49°, 22.88°, and 34.34° are related to cotton fibers, which correspond to the crystal plane diffraction of cellulose (101), (002), and (040) [30,31]. In addition, in Figure 6b–e, there are 7 other obvious sharp peaks and 2 weak diffraction peaks. The diffraction peaks of 31.78°, 34.43°, 36.26°, 47.55°, 56.61°, 62.87°, 66.40°, 67.97°, and 69.11° correspond to the crystal planes of ZnO (100), (002), (101), (102), (110), (103), (200), (112), and (201), respectively [32], showing that the ZnO NPs were loaded on the cotton fabric. The average crystallite size was determined by Scherrer's equation (Equation 1), the calculated crystallite size was 21.2 nm, and the size of the ZnO NPs loaded on the fabric became larger, the reason being that during the transfer of ZnO NPs from the treated solution to the fabrics, the treatments such as rolling, drying, and curing led to an agglomeration of the particles [19].

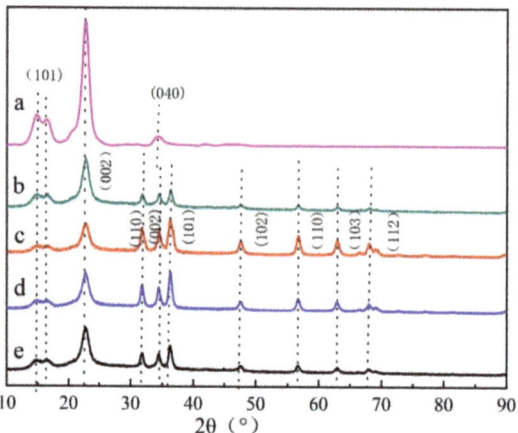

Figure 6. XRD patterns of raw cotton (**a**), ZnO/cotton (**b**), OA-ZnO/cotton (**c**), TA-ZnO/cotton (**d**), and SA-ZnO/cotton (**e**).

3.5. Optical Performance Analysis

The UV-vis diffuse reflectance spectrum and the relationship curve between $(\alpha h\nu)^{1/2}$ and the hν of the cotton fabrics loaded with ZnO NPs are given in Figure 7.

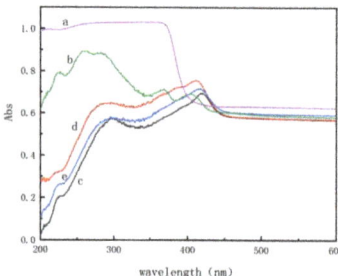 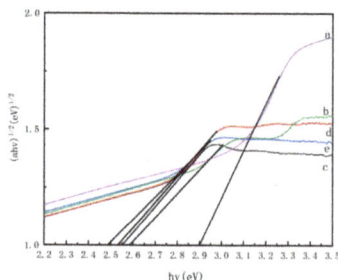

Figure 7. UV-vis diffuse reflectance spectra (**left**), $(Ah\nu)^{1/2}$ and hν relationship curves (**right**): Raw cotton (**a**), ZnO/cotton (**b**), OA-ZnO/cotton (**c**), TA-ZnO/cotton (**d**), and SA-ZnO/cotton (**e**).

It can be seen from Figure 7 left that in the ultraviolet region all fabrics show certain UV absorption properties, and the absorbance of ZnO/cotton was greater than that of the cross-linked cotton fabrics. This may be because the surface of ZnO/cotton adsorbs more ZnO NPs, and ZnO NPs have stronger absorption of ultraviolet light. While ZnO NPs of the cross-linked cotton fabrics were mainly fixed under the cross-linked film, the film affects its UV absorption intensity. However, in the visible light region, the absorbance of the cross-linked fabrics was greater than that of ZnO/cotton, which may be because the cross-linking film on the surface of the fabric also partially absorbs visible light. ZnO NPs are used as a crystalline semiconductor. The light absorption near the band edge follows the formula $\alpha h\nu = A(h\nu-E_g)^{n/2}$, where α, ν, E_g, and A are absorption coefficient, optical frequency, bandgap energy, and constant, respectively [33], where n depends on the characteristics of semiconductor electronic transitions, that is direct transition $n = 1$ and indirect transition $n = 4$ for ZnO, $n = 4$. The relationship between $(\alpha h\nu)^{1/2}$ and photon energy (hν) is shown on the right in Figure 7, where the intercept of the x-axis tangent is the forbidden bandwidth. The narrower the forbidden bandwidth of a semiconductor, the higher its light utilization. The right of Figure 7 shows that the bandgap width of ZnO prepared with honeysuckle extract was 2.90 eV, and ZnO/cotton was 2.57 eV. The bandgap of ZnO/cotton was narrower than that of ZnO NPs. This may be because cotton fabric also has a certain absorption effect on ultraviolet light, which increases its light utilization rate. Furthermore, the bandgap widths of OA-ZnO/cotton, TA-ZnO/cotton, and SA- ZnO/cotton were 2.55 eV, 2.54 eV and 2.48 eV, respectively. The bandgap widths of cross-linked cotton fabrics were narrower than that of the ZnO/cotton, indicating that cross-linking agents increased the utilization rate of cotton fabric to light.

3.6. Self-Cleaning Performance Analysis

The degradation rate of cotton fabrics after finishing with ZnO NPs was calculated by K/S before and after light exposure. The cotton fabrics stained with pollutants were placed under simulated sunlight, and the K/S values were measured at intervals of 1 h for a total of 4 h of irradiation. According to Formula (2), the degradation rates of cotton fabric under different times of light exposure were calculated, and the results are shown in Figure 8.

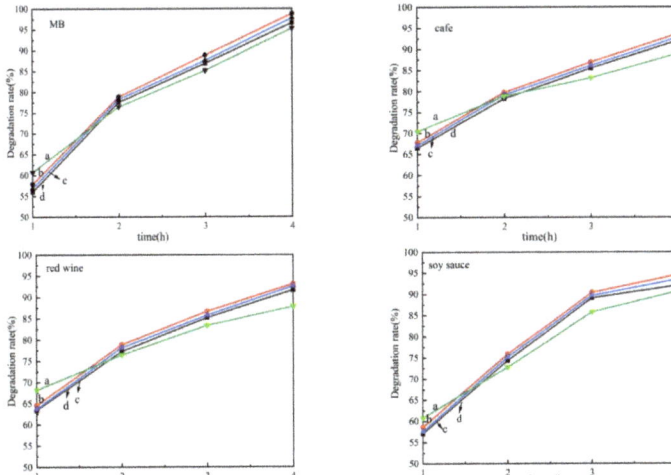

Figure 8. Degradation rates of ZnO/cotton (**a**), OA-ZnO/cotton (**b**), TA-ZnO/cotton (**c**), and SA-ZnO/cotton (**d**) to MB, coffee, red wine, and soy sauce under different irradiation times.

Figure 8 shows that all samples have a self-cleaning effect on methylene blue, coffee, red wine, and soy sauce. The degradation rate of ZnO/cotton fabrics at the initial stage of illumination is higher than that of cross linked cotton fabrics. This may be because most of the target pollutants were adsorbed on the ZnO/cotton surface during the initial stage of illumination. There were moreZnO NPs on the surface of ZnO/cotton, and a large number of free radicals generated by photoexcitation could contact the target pollutants and decompose them, so the degradation efficiency was higher, while for the cross-linked fabric, most of the target pollutants were adsorbed on the surface of the membrane during the initial stage of illumination. A few ZnO NPs were distributed on the surface of the membrane, so only fewer free radicals could directly contact the target pollutants, resulting in lower degradation efficiency. With the prolonged light time, the target pollutant diffused underneath the cross-linked membrane, at which time more free radicals could contact the pollutant and decompose it. Therefore, in the stage after contaminant diffusion has ended, the cross-linked cotton fabric had a high degradation rate of the target pollutant. In conclusion, improving the utilization rates of light of the cross-linked fabrics loaded with ZnO NPs led to increased degradation rates.

To evaluate the washing resistance of the cotton fabrics loaded with ZnO NPs, the SEM images and Znelement content of the fabrics by ICP-AES after 10 cycles of washing were tested, as shown in Figure 9.

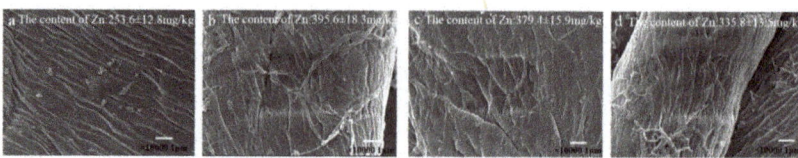

Figure 9. SEM images and Zn element contents of ZnO/cotton (**a**), OA-ZnO/cotton (**b**), TA-ZnO/cotton (**c**) and SA-ZnO/cotton (**d**) after 10 cycles of washing.

It can be seen from Figure 9 that the number of particles adsorbed on the ZnO/cotton surface obviously decreased after 10 cycles of washings, and the size of the particles became significantly smaller in contrast with the unwashed ZnO/cotton (Figure 4b), indicating that during the washing process, the larger ZnO NPs were likely to fall off the surface of the fabric, and smaller particles have relatively firm linkage to the fabric and did not easily fall off the surface of the fabric. Moreover, the surface of cross-linked fabrics still has only a few particles, but the surface of the fabrics became rough. This may be because the continuous washing caused the membrane structure on the surface of the cotton fabrics to be damaged.

It can be seen from the ICP data that after 10 cycles of washing, the content of Zn element in the ZnO/cotton decreased from 369.4 ± 7.6 mg/kg to 253.6 ± 12.8 mg/kg. This is because ZnO NPs were adsorbed by ZnO/cottons via weak intermolecular forces, so they tended to fall off during washing. The content of a Zn element of OA-ZnO/cotton, TA-ZnO/cotton, and SA-ZnO/cotton decreased from 546.4 ± 12.6 mg/kg, 487.8 ± 5.6 mg/kg, and 432.0 ± 8.9 mg/kg to 395.6 ± 18.3 mg/kg, 379.4 ± 15.9 mg/kg and 335.8 ± 13.5 mg/kg, respectively, indicating that the washing durability of the cross-linked cotton fabrics was greatly improved, possibly because the positively charged Zn^{2+} may be attracted by the negative charge of the carboxylate anion of the crosslinking agent [33]. The strong electrostatic attraction enforced the washing resistance of the fabric. Furthermore, the washing resistance of OA-ZnO/cotton was lower than that of TA-ZnO/cotton and SA-ZnO/cotton because the water contact angle (35.5°) of OA-ZnO/cotton was lower than that of TA-ZnO/cotton (46.34) and SA-ZnO/cotton (41.25°). Thus, the surface energy of OA-ZnO/cotton was higher, and the ZnO NPs were more likely to fall off during washing [28].

The degradation rates of the cotton fabrics after repeated washing and lighting for different cycles (0–10) were tested. The results are shown in Figure 10.

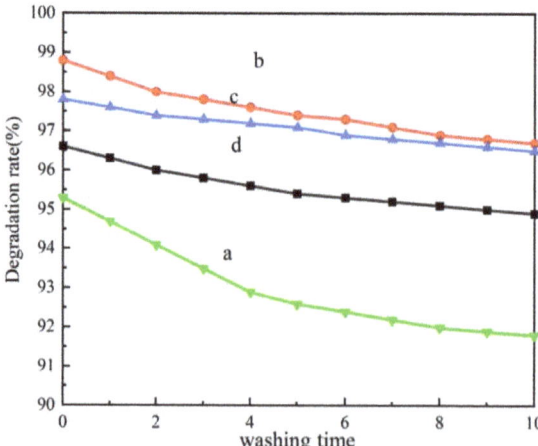

Figure 10. Degradation rates of methylene blue in ZnO/cotton (**a**), OA-ZnO/cotton (**b**), TA-ZnO/cotton (**c**) and SA-ZnO/cotton (**d**) at different washing cycles.

Figure 10 shows that after 10 cycles washes, the degradation rate of ZnO/cotton decreased more than OA-ZnO/cotton, TA-ZnO/cotton, and SA-ZnO/cotton. In addition, during the washing process of 1–4 cycles, the degradation rate of methylene blue by ZnO-cotton decreased greatly. This is because the larger particles of ZnO NPs adsorbed on the cotton fabric fall off the fabric first, resulting in a significant decrease in the degradation rate of the fabric. In the next few washing processes, the degradation rate decreased slowly because the retained small-size particles have relatively strong adsorption with the fabric and were not easy to fall off. However, the degradation rate of cross-linked fabrics decreased gently throughout the washing.

3.7. UV and Wrinkle Resistance of Fabric Analysis

The anti-ultraviolet and wrinkle resistance performances of the cotton fabrics were analyzed, and the results are given in Table 1. It can be seen from Table 1 that, compared with the raw cotton, the average UV transmittance of the UVA (320–400 nm) and UVB (280–320 nm) of ZnO/cotton, OA-ZnO/cotton, TA-ZnO/cotton, and SA-ZnO/cotton was significantly reduced, and the UPF values were significantly increased. It shows that the ZnO NPs loaded in the cotton fabrics greatly improved their UV resistance. This is because ZnO NPs can absorb photons with bandgap energy greater than 2.90 eV, and the photon energy in the ultraviolet region meets this requirement. After ZnO NPs absorb ultraviolet light energy, it not only produces oxidative degradation but also greatly reduces the ultraviolet transmission rate of the fabrics. In addition, the reflection and scattering of ultraviolet rays through the interface formed by the NPs and the fabric were also strengthened, thereby enhancing the shielding effect of the cotton fabrics to ultraviolet rays and improving its anti-ultraviolet performance [19]. According to the requirements of GB/T18830-2009, when the UPF value of the textile is more than 40, and its UVA transmittance is less than 5%, it is considered as an anti-ultraviolet product with a mark of 40+. So ZnO/cotton, OA-ZnO/cotton, TA-ZnO/cotton, and SA-ZnO/cotton can be called anti-ultraviolet textiles.

Table 1. UV and wrinkle resistance of the Raw Cotton, ZnO/Cotton, SA-ZnO/Cotton, OA-ZnO/cotton, and TA-ZnO/cotton.

Sample	Average UV Transmittance (%)		UPF	WRA(W + F(°))
	UVA	UVB		
Raw Cotton	29.98	15.65	5.36	120.1
ZnO/Cotton	1.77	1.49	50+	151.6
OA-ZnO/cotton	1.50	1.24	50+	171.6
TA-ZnO/cotton	1.30	1.02	50+	193.2
SA-ZnO/cotton	1.27	0.92	50+	225.7

In addition, compared with raw cotton, the wrinkle recovery angles of cotton fabrics loaded with ZnO NPs were improved. The wrinkle recovery angle of ZnO/cotton was slightly increased by 26.23%. However, the wrinkle recovery angles of OA-ZnO/cotton, TA-ZnO/cotton, and SA-ZnO/cotton were considerably increased by 87.93%, 51.21%, and 60.87%, respectively. This is due to the cross-linking of OA, SA, and TA with cotton fabrics effectively preventing the slippage of the cellulose macromolecular chain so that the cross-linked fabric has better resistance to deformation.

4. Conclusions

In this paper, honeysuckle extract was used to prepare ZnO NPs with an average particle size of 15.3 nm. Three dicarboxylic acids of OA, TA, and SA were adopted to cross-link cotton fabrics. Then ZnO NPs were applied to the cotton fabrics before and after cross-linking by the padding method to obtain the self-cleaning ZnO/cotton, OA-ZnO/cotton, TA-ZnO/cotton, and SA-ZnO/cotton. Through structure characterization and performance analysis of samples, we found that all the cotton fabrics finished with ZnO NPs have a high degradation ability to methylene blue, coffee, red wine, and soy sauce. Compared with the ZnO/cotton, the bandgap width of the cross-linked fabrics became narrower, the utilization rate to light was improved, and better self-cleaning performance was exhibited. A film formed on the surface of the cross-linked cotton fabrics, and ZnO NPs were mainly fixed under the film through strong electrostatic force, while the ZnO NPs on ZnO/cotton was adsorbed on the surface of the fabric through relatively weak van der Waals force, so the cross-linked cotton fabrics had better washing resistance. Moreover, for the cross-linked cotton fabrics, the adsorption amount of ZnO NPs of OA-ZnO/cotton was highest, the washing resistance of TA-ZnO/cotton was best, and after cross-linkage, the UV and wrinkle resistance of cotton fabrics was greatly improved.

Author Contributions: X.J., H.L. and J.Y. designed the experiments. X.J. and Y.Q. performed the experiments. X.J. analyzed the data. X.J. and Y.Q. wrote the paper. All authors have read and agreed to the published version of the manuscript.

Funding: This work was supported by the project of "Five Strategies" of Education Service of Industrial Technology Research Institute of Liaoning Universities (No. 2018LY027).

Institutional Review Board Statement: Not applicable.

Informed Consent Statement: Not applicable.

Data Availability Statement: The data presented in this study are available on request from the corresponding author.

Conflicts of Interest: The authors declare no conflict of interest.

References

1. Saxena, A.; Tripathi, R.M.; Singh, R.P. Biological synthesis of silver nanoparticles by using onion (*Allium cepa*) extract and their antibacterial activity. *Dig. J. Nanomater. Biostruct.* **2010**, *5*, 427–432.
2. Cruz, D.; Fale, P.L.; Mourato, A.; Vaz, P.D.; Luisa, S.M.; Lino, A.R.L. Preparation and physicochemical characterization of Ag nanoparticles biosynthesized by *Lippia citriodora* (lemon verbena). *Colloid Surf. B* **2010**, *81*, 67–73. [CrossRef] [PubMed]
3. Gunalan, S.; Sivaraj, R.; Rajendran, V. Green synthesized ZnO nanoparticles against bacterial and fungal pathogens. *Prog. Nat. Sci. Mater. Int.* **2012**, *22*, 693–700. [CrossRef]
4. Khun, K.; Ibupoto, Z.H.; Chey, C.O.; Lu, J.; Nur, O.; Willander, M. Comparative study of ZnO nanorods and thin films for chemical and biosensing applications and the development of ZnO nanorods based potentiometric strontium ion sensor. *Appl. Surf. Sci.* **2013**, *268*, 37–43. [CrossRef]
5. Vanaja, A.; Srinivisa Rao, K. Effect of Co Doping on Structural and Optical Properties of Zinc Oxide Nanoparticles Synthesized by Sol-Gel Method. *Adv. Nanopart.* **2016**, *5*, 83–89. [CrossRef]
6. Tang, E.; Tian, B.Y.; Zheng, E.; Fu, C.Y.; Cheng, G.X. Preparation of zinc oxide nanoparticle via uniform precipitation method and its surface modification by methacryloxypropyltrimethoxysilane. *Chem. Eng. Commun.* **2008**, *195*, 479–491. [CrossRef]
7. Sangeetha, G.; Rajeshwari, S.; Venckatesh, R. Green synthesis of zinc oxide nanoparticles by aloe barbadensis miller leaf extract: Structure and optical properties. *Mater. Res. Bull.* **2011**, *46*, 2560–2566. [CrossRef]
8. Taşdemir, A.; Aydin, R.; Akkaya, A.; Akman, N.; Altınay, Y.; Çetin, H.; Şahin, B.; Uzun, A.; Ayyıldız, E. A green approach for the preparation of nanostructured zinc oxide: Characterization and promising antibacterial behaviour. *Ceram. Int.* **2021**, *47*, 19362–19373. [CrossRef]
9. Wang, L.; Chang, L.X.; Zhao, B.; Yuan, Z.Y.; Shao, G.; Zheng, W.J. Systematic investigation on morphologies, forming mechanism, photocatalytic and photoluminescent properties of ZnO nanostructures constructed in ionic liquids. *Inorg. Chem.* **2008**, *47*, 1443–1452. [CrossRef]
10. Deng, Z.W.; Chen, M.; Gu, G.X.; Wu, L.M. A facile method to fabricate ZnO hollow spheres and their photocatalytic property. *J. Phys. Chem. B* **2008**, *112*, 16–22. [CrossRef]
11. Zuo, D.Y.; Li, G.Q.; Ling, Y.L.; Cheng, S.J.; Xu, J.; Zhang, H.W. Durable UV-blocking Property of Cotton Fabrics with Nanocomposite Coating Based on Graphene Oxide/ZnO Quantum Dot via Water-based Self-assembly. *Fibers Polym.* **2021**, *22*, 1837–1843. [CrossRef]
12. Amani, A.; Montazer, M.; Mahmoudirad, M. Synthesis of applicable hydrogel corn silk/ZnO nanocomposites on polyester fabric with antimicrobial properties and low cytotoxicity. *Int. J. Biol. Macromol.* **2019**, *123*, 1079–1090. [CrossRef]
13. Ankamwar, B.; Chahudhary, M.; Sastry, M. Gold nanotriangles biologically synthesized using tamarind leaf extract and potential application in vapor sensing. *Synth. React. Inorg. Met.-Org. Nano-Met. Chem.* **2005**, *35*, 19–26. [CrossRef]
14. Hossain, M.F.; Asaduzzaman, M.U. Optimum Concentration of Cross-Linking Agent for Crease-Resistant Finishing Using DMDHEU on Knit and Woven Fabric in the Field of Ease Care Durable Press. *J. Text. Sci.* **2016**, *5*, 74–81.
15. Cai, X.Y.; Li, H.; Zhang, L.; Yan, J. Dyeing Property Improvement of Madder with Polycarboxylic Acid for Cotton. *Polymers* **2021**, *13*, 3289. [CrossRef]
16. Kumar, R.R.; Yazhini, K.B.; Prabu, H.G.; Zhou, Q.X. Polyfunctional Application on Modified Cotton Fabric. *Natl. Acad. Sci. Lett.* **2019**, *42*, 475–478. [CrossRef]

17. Dong, Y.M.; Liu, X.Y.; Liu, J.; Yan, Y.; Liu, X.; Wang, K.; Li, J. Evaluation of anti-mold, termite resistance and physical-mechanical properties of bamboo cross-linking modified by polycarboxylic acids. *Constr. Build. Mater.* **2021**, *272*, 121953. [CrossRef]
18. Majid, M.; Alimohammadi, F.; Shamei, A.; Rahimi, M.K. Durable antibacterial and cross-linking cotton with colloidal silver nanoparticles and butane tetracarboxylic acid without yellowing. *Colloid Surf. B* **2012**, *89*, 196–202.
19. Zhu, J.L.; Li, H.; Wang, Y.; Wang, Y.S.; Yan, J. Preparation of Ag NPs and Its Multifunctional Finishing for Cotton Fabric. *Polymers* **2021**, *13*, 1338. [CrossRef]
20. Mirjalili, M.; Karimi, L. Photocatalytic Degradation of Synthesized Colorant Stains on Cotton Fabric Coated with Nano TiO$_2$. *J. Fiber Bioeng. Inform.* **2010**, *3*, 208–215. [CrossRef]
21. Tran, V.H.T.; Lee, B.K. Development of multifunctional self-cleaning and UV blocking cotton fabric with modification of photoactive ZnO coating via microwave method. *J. Photochem. Photobiol. A* **2017**, *338*, 13–22. [CrossRef]
22. Uribe López, M.C.; Hidalgo López, M.C.; LópezGonzález, R.; Frías Márquez, D.M.; Núñez Nogueira, G.; Hernández Castillo, D.; Alvarez Lemus, M.A. Photocatalytic activity of ZnO nanoparticles and the role of the synthesis method on their physical and chemical properties. *J. Photochem. Photobiol. A* **2021**, *404*, 112866. [CrossRef]
23. Peng, Y.G.; Ji, J.L.; Zhao, X.Y.; Wan, H.X.; Chen, D.J. Preparation of ZnOnanopowder by a novel ultrasound assisted non-hydrolytic sol–gel process and its application in photocatalytic degradation of C.I. Acid Red 249. *Powder Technol.* **2013**, *233*, 325–330. [CrossRef]
24. Cao, Y.; Tan, H.M. Structural characterization of cellulose with enzymatic treatment. *J. Mol. Struct.* **2004**, *705*, 189–193. [CrossRef]
25. Mahamuni, P.P.; Patil, P.M.; Dhanavade, M.J.; Badiger, M.; Shadija, P.G.; Lokhande, A.C. Synthesis and characterization of zinc oxide nanoparticles by using polyol chemistry for their antimicrobial and antibiofilm activity. *Biochem. Biophys. Rep.* **2018**, *17*, 71–80. [CrossRef] [PubMed]
26. Sarwar, N.; Ashraf, M.; Mohsin, M.; Rehman, M.; Younus, A.; Javid, A.; Iqbal, K.; Riaz, S. Multifunctional Formaldehyde Free Finishing of Cotton by Using Metal Oxide Nanoparticles and Ecofriendly Cross-Linkers. *Fibers Polym.* **2019**, *20*, 2326–2333. [CrossRef]
27. Cornils, B.; Lappe, P. Dicarboxylic Acids, Aliphatic. In *Ullmann's Encyclopedia of Industrial Chemistry*; Wiley-VCH: Weinheim, Germany, 2006.
28. Khajavi, R.; Berendjchi, A. Effect of Dicarboxylic Acid Chain Length on the Self-Cleaning Property of Nano-TiO$_2$-Coated Cotton Fabrics. *ACS Appl. Mater. Interfaces* **2014**, *6*, 18795–18799. [CrossRef] [PubMed]
29. Karimi, L.; Mirjalili, M.; Yazdanshenas, M.E.; Nazari, A. Effect of Nano TiO$_2$ on Self-cleaning Property of Cross-linking Cotton Fabric with Succinic Acid under UV Irradiation. *Photochem. Photobiol.* **2010**, *86*, 1030–1037. [CrossRef] [PubMed]
30. Siami, A.M.; Majid, M. Clean Sono-synthesis of ZnO on Cotton/Nylon Fabric Using Dopamine: Photocatalytic, Hydrophilic, Antibacterial Features. *Fibers Polym.* **2021**, *22*, 97–108.
31. Hu, C.; Zhou, Y.Y.; Zhang, T.; Jiang, T.J.; Meng, C.; Zeng, G.S. Morphological, Thermal, Mechanical, and Optical Properties of Hybrid Nanocellulose Film Containing Cellulose Nanofiber and Cellulose Nanocrystals. *Fibers Polym.* **2021**, *22*, 2187–2193. [CrossRef]
32. Gokula, K.P.; Muthukumaran, S.; Raja, V. Structural, energy gap tuning, photoluminescence and magnetic properties of Sn-doped $Zn_{0.96}Ni_{0.04}O$ nanostructures. *J. Lumin.* **2021**, *238*, 118258.
33. Zhang, X.; Ai, Z.H.; Jia, F.J.; Jia, F.L.; Zhang, L.Z. Generalized One-Pot Synthesis, Characterization, and Photocatalytic Activity of Hierarchical BiOX (X = Cl, Br, I) Nanoplate Microspheres. *J. Phys. Chem. C* **2008**, *112*, 747–753. [CrossRef]

Article

Parametric Analyses of the Influence of Temperature, Load Duration, and Interlayer Thickness on a Laminated Glass Structure Exposed to Out-of-Plane Loading

Mirela Galić *, Gabrijela Grozdanić, Vladimir Divić and Pavao Marović

Faculty of Civil Engineering, Architecture and Geodesy, University of Split, 21000 Split, Croatia; gabrijela.grozdanic@gradst.hr (G.G.); vladimir.divic@gradst.hr (V.D.); marovic@gradst.hr (P.M.)
* Correspondence: mirela.galic@gradst.hr; Tel.: +385-21-303-334

Abstract: One of today's most-used glass products is a composite made of at least two glass panels connected with a soft polymeric interlayer—laminated glass. The mechanical properties of such elements are influenced by interlayer properties and the type of glass used. In this work, experimental and numerical analyses of laminated glass panels exposed to four-point bending are performed to observe and compare the stresses and displacements caused by different parameters, such as temperature, load duration, the thickness and type of the interlayers, as well as the symmetrical and nonsymmetrical disposition of the glass plates' thickness. The numerical analysis was verified by four-point bending experimental tests. After validation, a parametric study on these influences was performed. To obtain the relationship between the load duration, temperature, and thickness of the interlayer compared to the maximal displacement (as a measure of flexural stiffness) and tension stress in the bottom glass plate, an analytical polynomial of a sixth total order is proposed. Isosurfaces are created, showing the dependence of stresses and displacements on the specified parameters as well as clearly showing differences in the behavior of laminated glass panels for the same conditions but with different interlayers. Based on the findings of the parametric study, conclusions are derived about the flexural stiffness and stress distribution in two-plate laminated glass with PVB and ionoplast interlayers.

Keywords: laminated glass; interlayer; experimental test; numerical model; parametric analysis

Citation: Galić, M.; Grozdanić, G.; Divić, V.; Marović, P. Parametric Analyses of the Influence of Temperature, Load Duration, and Interlayer Thickness on a Laminated Glass Structure Exposed to Out-of-Plane Loading. *Crystals* 2022, 12, 838. https://doi.org/10.3390/cryst12060838

Academic Editor: Alessandro Chiasera

Received: 8 April 2022
Accepted: 11 June 2022
Published: 14 June 2022

Publisher's Note: MDPI stays neutral with regard to jurisdictional claims in published maps and institutional affiliations.

Copyright: © 2022 by the authors. Licensee MDPI, Basel, Switzerland. This article is an open access article distributed under the terms and conditions of the Creative Commons Attribution (CC BY) license (https://creativecommons.org/licenses/by/4.0/).

1. Introduction

Laminated glass consists of several glass plates connected by polymorphic intermediate layers. Today, facade systems are made of this material, areas above valuable archaeological sites are paved with it, and attractive pedestrian bridges as well as stairs are made with it. Interest in the use of laminated glass in construction has increased due to the fact that laminated glass retains a particularly interesting transparency compared to monolithic glass, and its advantages over monolithic glass include greater impact resistance, improved sound resistance, and better thermal insulation inside a building. The most important advantage of laminated glass over monolithic glass is its improved safety, which is most visible in the case of breakages of the structure, i.e., if the glass surface is broken into fragments, the interlayer will preserve them together and prevent complete disintegration, avoiding destruction or injuries. Despite the growing use of these types of glass structures, the norms for their calculation are not yet fully defined, nor are the national supplements of individual countries. Therefore, it is important to experimentally and numerically test such structures that will contribute to the improvement and completion of engineering calculations. Given the fact that such elements are exposed to different atmospheric and environmental conditions, it is important to examine how they behave under these influences. One of these is the determination of the performance of laminated glass at different temperatures in addition to its response to load over time due to temperature changes.

Since we are talking about the composite material of glass and interlayers, it is necessary to include the action and change in the performance of each material separately, but also to consider the behavior of the whole.

Therefore, the most important characteristics of glass as a material and of the interlayers will be shown below, with reference to previous research on both.

1.1. Glass as Material

SiO_2-Na_2O-CaO (with additional ingredients) is the most commonly used type of glass, called soda–lime glass (referred to in the text as "glass"). Soda-lime glass, produced by the float process, can be used without any additional treatment if there are no additional demands for increased safety, strength, design, etc. In the production process, a molten tin (tin bath) is a base for cooling and transporting a molten glass, consequently producing a flat surface that does not require any postprocessing treatment. After acquiring a solid state in a further part of the production process, the glass ribbon is heated and slowly cooled to release residual stresses (annealing).

The mechanical properties of glass and the characteristic bending strengths according to EN16612 [1] are presented in Tables 1 and 2, respectively. By adding different additional ingredients to the chemical composition, it is possible to impact some glass properties, such as heat resistance, color, etc. Except for changes in ingredients, there are physical treatments that can improve glass mechanical properties. One of the most-used processes is tempering, a process that can be accomplished by a chemical treatment or by heating and quickly cooling glass panels. The final result of both processes is prestressing, which results in an increased tensile strength of the glass panels. Besides higher strength, glass develops the ability of safe breakage. The difference between the fracture pattern of the annealed and tempered glass is presented in Figure 1. Annealed glass produces large and sharp pieces that can cause injuries, while tempered glass develops small, cubical, and harmless pieces. Heat-strengthened glass is tempered glass but with a lower level of prestressing [2]. The glass modulus is not affected by additional processing (tempering, etc.) or atmospheric conditions. Besides the glass type, the loading also has a significant impact on the breakage pattern. For static loading, the cracks are primarily transversal and straight, and the total length of the cracks is smaller than those that are circular-shaped and produced by impact loading [3–6]. The mechanical characteristics of glass are not affected by changes in the temperature in the interval of the ambient temperatures.

Table 1. Basic mechanical properties of glass according to EN16612 [1].

Properties	Middle Value	Interval
Glass density	ρ = 2500 kg/m^3	2250–2750 kg/m^3
Young's modulus	E = 70,000 MPa	63,000–77,000 MPa
Poisson number	μ = 0.23	0.20–0.25

Table 2. Characteristic bending strength of each type of glass according to EN16612 [1].

Annealed Glass/Float Glass	Heat-Strengthened Glass (HSG)	Thermally Toughened Glass (TTG)
45 N/mm^2	70 N/mm^2	120 N/mm^2

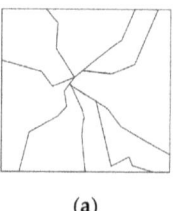 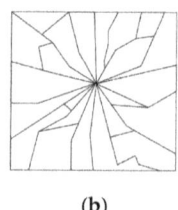 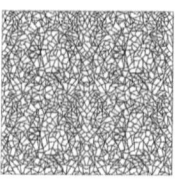

(a)   (b)   (c)

Figure 1. Breakage pattern for each type of glass: (**a**) annealed glass/float glass; (**b**) heat-strengthened glass; (**c**) thermally toughened glass [2].

1.2. Laminated Glass—Type of Interlayers

Laminated glass can be produced from any type of glass by coupling at least two glass panels with an interlayer, providing a better load capacity and postbreakage safety by ensuring the integrity of glass during a breakage. Interlayers in laminated glass are usually transparent polymer materials with large variations in their mechanical properties depending on temperature, load duration, moisture, etc., and they ensure coupled behavior and a postbreakage capacity [3–8]. The mechanical behavior of laminated glass structures could be bounded between two limit states: one in which we neglect the influence of interlayers and observe laminated glass as two separate plates without friction, and a monolithic behavior where we observe a laminated glass panel with full shear transfer through an interlayer [9,10]. The positioning of a structure within the presented limits depends on the type of loading and load duration, the mechanical characteristics of the interlayer, and the type of boundary conditions. For high-velocity loads (impact), the interlayer mostly behaves very stiffly, and it is possible to observe the panel as monolithic, while at static loading we emphasize the influence of the viscoelastic character of the interlayer.

Once a plate is broken, fragments remain adhered to the interlayer. In a tensile zone, that adhesion does not provide any specific benefit. However, in the compression zone, the remaining glass fragments can create additional bearing capacity while making contact with each other, and by doing so, transfer compression stresses.

1.3. Influence of Interlayers' Behavior on the Capacity of Laminated Glass Plates

The most commonly used interlayers in laminated glass structures are PVB (polyvinyl butyral), EVA (ethylene vinyl acetate), and ionoplast. In a static analysis, the mechanical properties of interlayers are primarily affected by temperature conditions, moisture, and load duration. Therefore, the researchers were engaged in the experimental and computer testing of samples of these materials in addition to defining parameters to describe the behavior of these materials. Studies in the literature [11] have shown that the mechanical properties of polymer interlayers show dependence on multiple external influences, such as current temperature, heat cycles, or exposure to moisture. The influence of temperature as a parameter in the analysis of influences on the relaxation behavior of polymers was analyzed in [12], in which it was found that an increase in the temperature reduces the relaxation time of the polymer. This property was included in the model as a variable of a function of time [13]. Taking all of the above characteristics into consideration, the material parameters of the generalized Maxwell chain model for two types of interlayers, such as PVB-based (TROSIFOL® BG R20 from Kuraray, Tokyo, Japan) and EVA-based EVALAM® 80–120 (from Evalam, Coruna, Spain), were defined, which enables the description of their time- and temperature-dependent behavior [14]. Static shear experimental testing [15] and long-term testing [16] also showed shear modulus degradation for temperature increases and for longer load durations.

For numerical simulations of laminated glass behavior, different models of interlayers are used. The chosen model of interlayer material depends on the type of loading: hyperelastic models [5] and rate-dependent hyperelastic models [7] are often used for the dynamic type of loading, while some authors consider elastoplastic models for high strain rates [17] and nonlinear elastic hardening materials [18] for static loading.

The material specifications of interlayers are summarized in Table 3, and graphs with Young's moduli are presented in Figure 2. All information regarding ionoplast and PVB interlayers is taken from the available technical documentation [19,20]. The thickness of an interlayer is mainly in the range of 0.36 to 2.28 mm; it differs depending on the type of interlayer and the physical requirements.

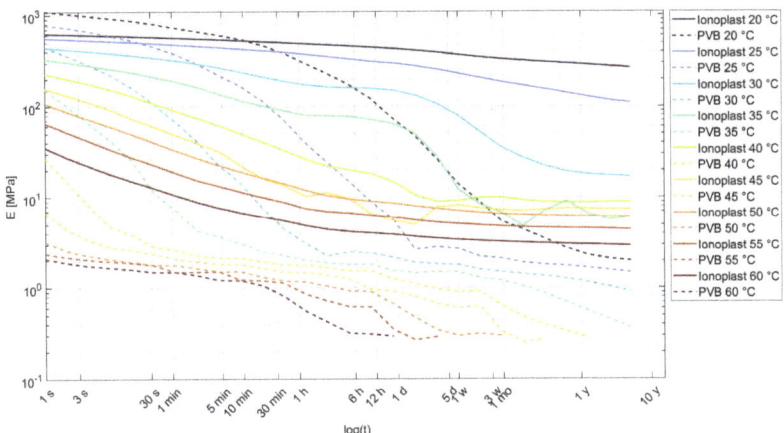

Figure 2. Comparison of Young's modulus, E (MPa), degradation for the ionoplast and PVB interlayers in dependence on the load duration log, (t(s)), at different temperatures, T (°C) [19,20].

PVB (polyvinyl butyral) is a synthetic polymer of the polyvinyl acetate family, one of the most-used materials for coupling glass panels. The mechanical properties of this material show a significant dependence on temperature, load duration [17,21,22], and moisture [23]. There are numerous types of PVB interlayers used for different primary purposes (acoustic, structural, and solar). These types of interlayers have different material properties, resulting in different stiffnesses and different glass transition temperatures (T_g) [24]. Here, we will observe the structural behavior of PVB. The mechanical properties (shear modulus) of PVB interlayers show a discrepancy when comparing specimens produced from PVB before the autoclave process and those who are embedded inside laminated glass with the autoclave process [25]. Since heat and pressure from the autoclave process have an influence on this material, it is more preferable to observe PVB characteristics while embedded in laminated glass. PVB interlayers show significant degradation of shear transfer at increased temperatures.

EVA (ethylene vinyl acetate) is a copolymer interlayer material, a moisture-resistant interlayer [23] that is often used for specific purposes, such as photovoltaic cells or if a colored design is required. Except for some experiments, which provide specific types of results, there is not much available information about the material characteristics of EVA interlayers. The production of laminated glass with this interlayer is less demanding and does not require an autoclave process [24]. Tests on small-scale samples of laminated glass with PVB and EVA interlayers from the literature (see [14]), in a dynamic single-lap shear test and dynamic torsion tests, show differences between these two interlayers. For samples tested at different frequencies (dynamic tests) and temperatures, PVB showed a generally stiffer response compared to EVA, but that was valid only for temperatures under 40 °C. Additionally, at a significantly lower temperature (around −20 °C), where other interlayers have almost brittle behavior, EVA interlayers show better impact resistance (penetration resistance) due to their low glass-transition temperatures [24]. On the other hand, four-point bending tests at room temperature [26] showed similar behavior of a laminated glass panel with an EVA interlayer and PVB interlayer. EVA interlayers have thermorheologically complex behavior [27]. By testing samples of EVA interlayers, authors [28] describe behavior with two types of constitutive models regarding the type of loading. At a small strain, EVA is described with time-dependent behavior (linear viscoelasticity), and for large deformations, where nonlinear stress–strain behavior occurs, material behavior is described with a hyperelastic model. These models are derived from dynamic mechanical thermal analysis (DMTA), uniaxial tests, and biaxial tests, and they can be used for modeling the postbreakage capacity of laminated elements.

In [29], the authors tested samples of different types of PVB and EVA interlayers loaded in static single-lap shear tests, and in [30] PVB samples were tested in double-lap long-term tests. In [29], the temperature and strain rate are varied, and a different humidity and temperature are observed in [30]. Both tests confirmed the significant dependence of interlayer characteristics on temperature, as well as their minor sensitivity to humidity. These shear tests could be characterized as being more appropriate for describing interlayer behavior in laminated glass panels in an unfractured state. In [31], the authors first presented an overview of the generally used methods for determining polymer thermoviscoelastic behavior, and thereafter presented the results of dynamic-torsion cyclic tests on small samples of laminated glass. In these types of tests, laminated glass samples are glued for rheometers, which can affect the final results.

When observing all of the mentioned tests, one can find that some tests for the determination of interlayer mechanical characteristics are performed on samples of interlayer materials only (no glass panels involved), and others are performed on coupled elements with glass parts. Interlayers are not predetermined for usage as stand-only elements, and in laminated glass panels, they are dominantly loaded in shear. Tests on raw materials should always be validated with large-scale tests on laminated glass [32].

An ionoplast interlayer is an ionomer-based material that provides the highest level of structural performance [23], but it is also the most expensive one. It was originally developed for hurricane-resistant glass surfaces on building facades [6]. Since the degree of coupling glass plates in a laminated structure depends upon the shear stiffness of the polymeric interlayer, ionoplast is the best option when there are high demands for strength and resistance [33]. In four-point bending experimental tests [34], laminated glass with an ionoplast interlayer shows a higher initial failure load and ultimate load (providing better postbreakage capacity) than laminated glass with PVB and EVA interlayers.

Table 3. Basic mechanical properties of the interlayers [19,20,24].

	PVB (Structural)	EVA [34]	Ionoplast
Density	1070 kg/m^3	970 kg/m^3	950 kg/m^3
Poisson's ratio	0.476	0.32	0.458
Glass transition temperature [24]	12–25 °C	−28 °C	55 °C

Regarding everything that has been mentioned, in the numerical model we included the material model of the interlayers, which involves the degradation of Young's modulus, E, and the shear modulus, G, in dependence on the load duration log, (t(s)), at different temperatures, T (°C). These parameters are dominant in laminated glass structures' exposed bending.

2. Experimental Tests

The available data and the mentioned analyses suggest the need for a parametric analysis of the influence of temperature change on the behavior of laminated glass loaded outside the plane. In order to perform this analysis, it is necessary to verify the numerical model with an experimental test. The experimental test on laminated glass specimens was carried out in the Structural Laboratory of the Faculty of Civil Engineering, Architecture and Geodesy. Laminated glass specimens were exposed to four-point bending. Specimens were made out of two tempered glass panels with a thickness of 6 mm. Glass panels were connected with a 0.76 mm-thick Saflex DG41 PVB (EASTMAN, Kingsport, TN, USA; $E_{1min, 25\,°C} = 387$ MPa; $G_{1min, 25\,°C} = 131$ MPa; $\nu = 0.476$) interlayer. The span was 950 mm, while the width and length of the specimen were 330 mm and 1000 mm, respectively. Hard contact of the steel bearing and the glass panel was prevented with 0.1 mm-thick rubber protection. The specimens were bent uniformly, increasing the bending stress at a rate of 2.0 N/mm^2.s, until failure occurred, according to EN 1288-3 [35]. Specimens were exposed to the same load at the same temperature (25 °C) and moisture conditions (50%). All of the specimens were produced one year before the test, and they were stored in the same conditions.

The experiments were performed on a CONTROLS Automax Multitest (CONTROLS, Milan, Italy) loading device. The samples were placed on cylindrical supports and separated with rubber pads. For the data acquisition, a QUANTUM MX840B (HBM, Darmstadt, Germany) was employed, with a sampling frequency of 300 Hz. Force and displacement were measured with CONTROLS implemented acquisition. Additionally, the deflections were measured with six linear variable displacement transducers (LVDTs), which were arranged symmetrically on the specimen: two at the mid-span and four at the bearings. The applied force was controlled by a CONTROLS device at each step.

The test setup is presented in Figures 3 and 4. Strains at the upper and bottom glass plates at points A and B (marked on Figure 3) were measured with four strain gauges. In all of the specimens, fractures occurred on the bottom plate, while the upper plate remained undamaged; see Figure 5. After the initial cracking of the bottom plate, the load-bearing capacity is reduced. After complete cracking, the glass can no longer withstand tensile stresses, but glass fragments remain attached to the polymer interlayer. In this phase, the polymer can provide the tensile force together with the lower part of the upper plate to withstand bending moments. At this stage, the load capacity also depends on the size of the fragments, which is directly related to the type of glass. In our experimental tests, fracturing occurs at the bottom plate, and such a cracked panel remains loaded with 1 kN of force without a significant increase in deflection.

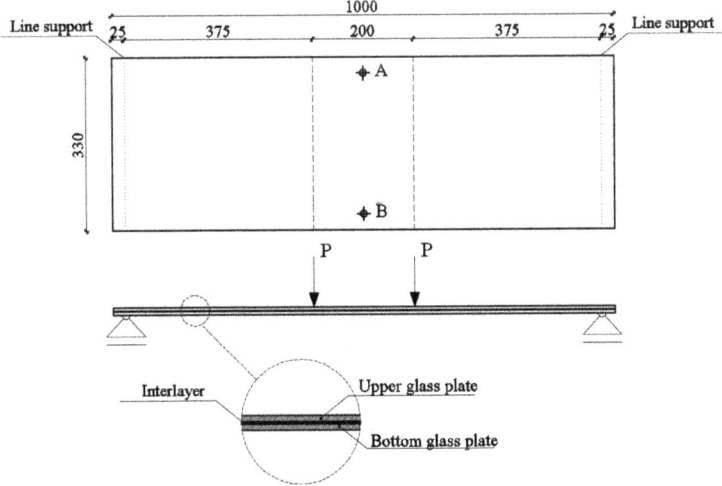

Figure 3. Test setup for the four-point bending test.

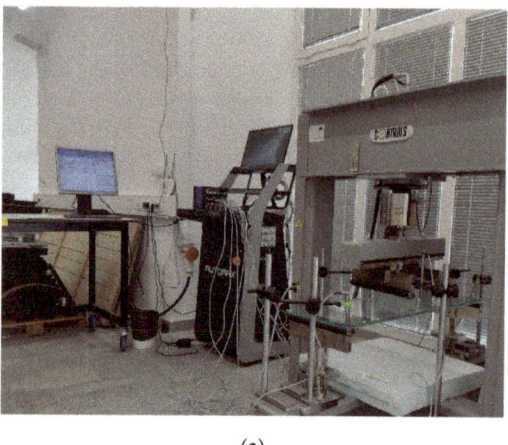
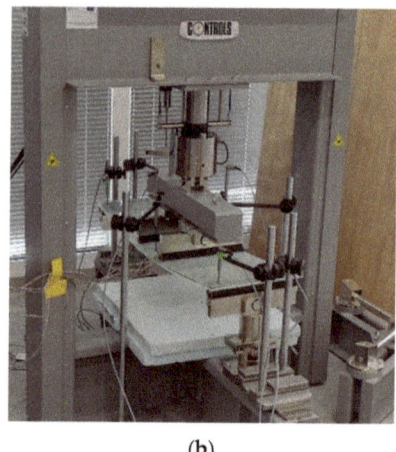

Figure 4. A view of the experiment: (**a**) test setup before loading; (**b**) breakage of bottom glass plate during the experiment.

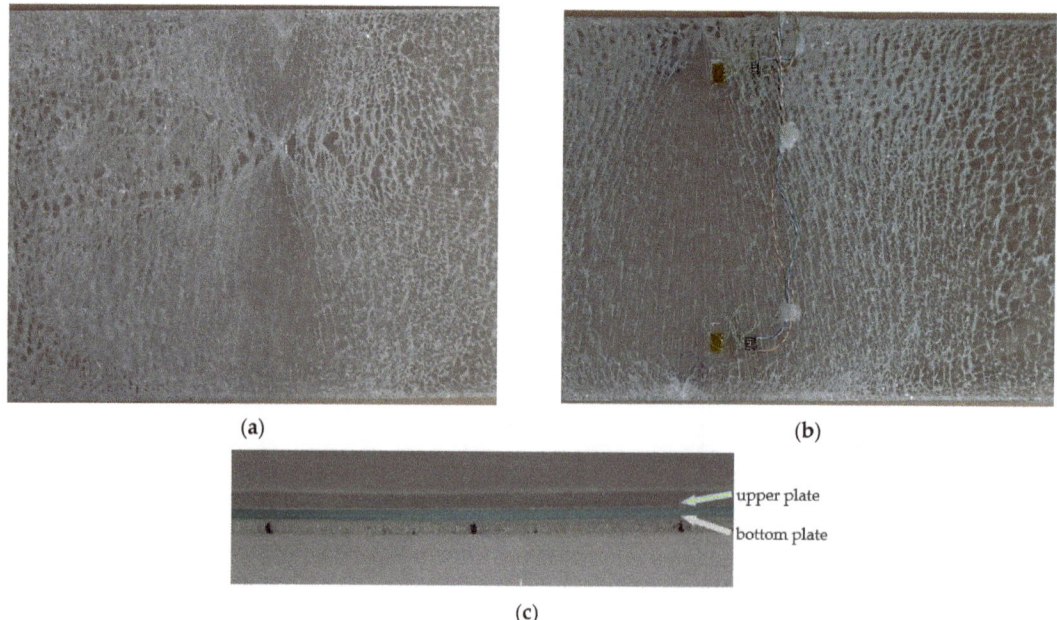

Figure 5. Photography of fracture pattern on specimens: (**a**) S1; (**b**) S3; and (**c**) side view on center of laminated glass panel.

The results for all of the tested specimens are presented in Figure 6 and compared to the four-point bending tests on glass elements of Pankhardt and Balázs [8]. In the graph, only the results of the laminated glass panel tests are presented. It is evident that the results at room temperature do not differ significantly. The contribution of the interlayer to the global behavior of the glass panels at room temperature is minimal. Another comparison curve is from the experimental tests provided by Serafinavicius et al. [6], where the authors tested laminated glass composed of two 6 mm thick panels with PVB (1.52 mm), EVA (0.89 mm), and SGP (1.52 mm) interlayers.

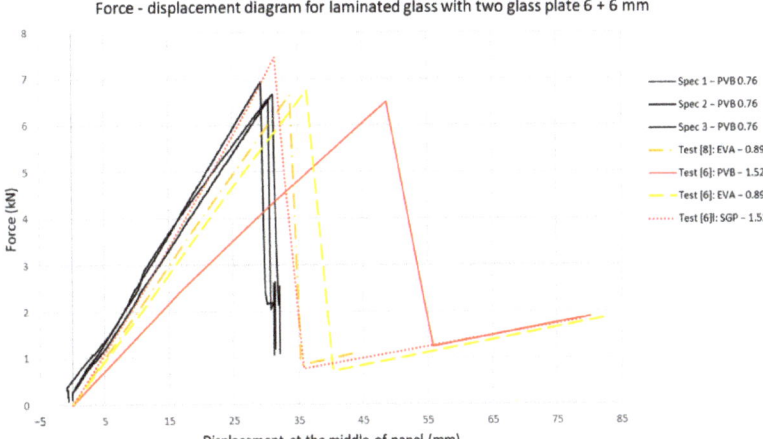

Figure 6. Results of the four-point bending test: force-displacement diagram compared with experimental results from the literature [6,8].

These results showed a good accordance, regardless of the differences in span and the slightly different thicknesses of the interlayers. It is visible that at room temperature and for similar thicknesses (0.89 mm and 0.76 mm), the EVA [6,8] and PVB interlayers provide approximately the same ultimate loading. As expected, the SentryGlas ionoplast interlayer, at the same thickness (1.52 mm) as PVB [6], shows a much stiffer response, and thus an increase in the ultimate strength. It is important to mention that we defined the term "ultimate strength" as a breakage of the bottom (tensile) glass plate, while the upper glass plate remains undamaged. The performed tests, as well as the results from [6], confirm that after the breakage of the bottom glass plate, the laminated glass panel provides bearing capacity. After breakage, the panels are unloaded, but a permanent deformation of 12 mm remains. The deformed and damaged specimens are removed and placed on a flat surface with the broken glass plate facing upwards (loaded with own weight). The permanent deformation remains even 3 months after the test, regardless of the slight tendency of a weight force to straighten the panel. This permanent deformation is a product of bulk volume increase in the breakage of a tempered glass panel, which as a consequence has a plastic deformation of the interlayer.

Although interlayers and glass panels have significantly different mechanical behavior, it is proven [36] that, with increasing interlayer thickness, the ultimate load of glass panels also increases.

In all of the specimens exposed to four-point bending, fractures occur in the zone of the maximum moment (between the applied forces) on tensile glass plates, which is typical for static loading. It is visible in [37] that dynamic loading fracturing mostly occurs at the contact surface.

3. Numerical Model

Based on geometry and the loading of the presented experimental tests, the numerical model is developed. The model is developed in ANSYS software, and it consists of two glass plates connected with an interlayer. The model is discretized with 3D solid elements, with minimal two-element through-thickness of the interlayer, and the element-size ratio for the glass panel is not bigger than 1.5 interlayer elements. Contact between the glass and interlayer is defined as an absolute bond, such that only interlayer shear deformation occurs in the middle, without any sliding. The glass panel is supported on two ends with a span of 95 cm. The load is placed at the same positions as in the four-point bending test, with two line loads at equal distances from center of plate, 10 cm on each side from the mid-span. The used material characteristics for the glass are shown in Tables 1 and 2, and those for PVB in Table 3, as well as Figure 2. Some basic information for the numerical model is presented in Table 4. For the validation of the experimental test, Young's modulus and Poisson's ratio are adopted according to the total test duration and test temperature ($E_{1min,25\,°C}$ = 387 MPa; $G_{1min,25\,°C}$ = 131 MPa; ν = 0.476). The shear modulus is calculated in dependence to Young's modulus and Poisson's ratio. The numerical model (with PVB) is validated with the presented experimental data for the first loading stage (fracture on the bottom glass plate). In Figure 7, a good coincidence in the numerical compared to the experimental results can be seen. Since the obtained numerical results showed a good accordance with the experimental results, this model was considered to be validated for further analyses. This all served as an introduction and validation for the following parametric analysis.

Table 4. Basic information about the finite-element model in ANSYS [38].

Material	Element Type	Material Model	Contact
Glass	Solid, full integration	Linear elastic (E = 70 GPa; ν = 0.23)	Bonded, no separation
Interlayer	Solid, full integration	Load duration and temperature dependent	

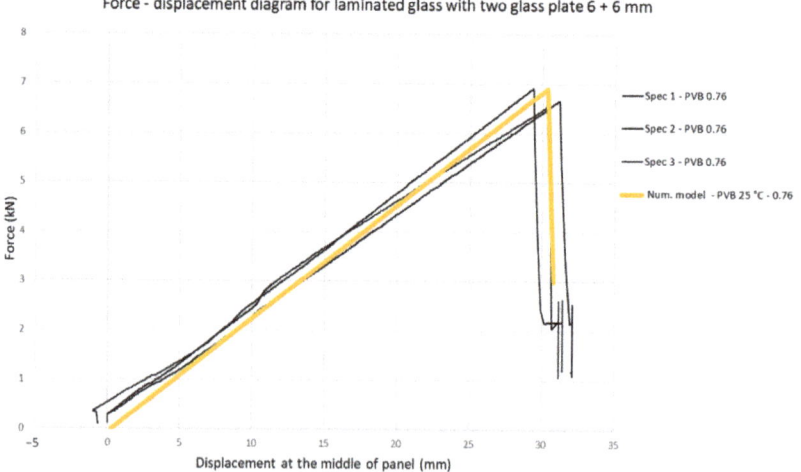

Figure 7. Force-displacement diagram: results of four-point bending from the numerical and experimental tests.

For the PVB material model in further analyses, different values of Young's modulus and the shear modulus, G, in dependence to the load duration and temperature are used (Figure 2). This is accepted because in observed bending tests, the panel is loaded until the fracturing of one plate occurs. For this type of loading, the interlayer is exposed to a small strain, lower than the failure strain of the glass that occurs on laminated glass elements [21]. The failure strain of the glass plate is approximately 0.167% for a glass strength of 120 MPa and an elastic modulus of 70 GPa. Since the interlayer is placed in the middle of the laminated glass element with a dominant bending deformation, it is exposed to a very small strain in the unfractured state.

Based on the validated numerical model, other models with different geometries (regarding the possible thickness of the glass and interlayer) and interlayers are developed. These models are used in the parametric study.

4. A Parametric Study of Interlayer Properties on Bearing Capacity of Laminated Glass Panels

For the parametric analysis, twelve different geometries are developed. All of the models have a span of 950 mm and a width of 330 mm. Each panel is loaded with a total force of 1 kN divided into two lines uniformly distributed with 15,151 N/mm'. The description of the geometries and interlayers is presented in Table 5. For the interlayer thicknesses of t = 1.52 mm and t = 2.28 mm, both PVB and ionoplast interlayers are used, while for lower thicknesses PVB is used in the models with t = 0.76 mm and ionoplast for the models with t = 0.89 mm. The main goal is to use the same total thickness of glass panels but nonsymmetrically placed in the tensile and compressive zones (glass thickness of 10 mm + 6 mm and 6 mm + 10 mm). With a fixed force, the temperature and load duration are varied, and the influence of those two parameters on stresses in the bottom glass plate as well as total deflection is analyzed for different geometries.

Table 5. Basic information about geometries of finite-element models in ANSYS [38].

Interlayer Thickness	Glass Thickness (Upper + Bottom Plates)	Interlayer
0.76 mm	8 mm + 8 mm	PVB
0.76 mm	10 mm + 6 mm	PVB
0.76 mm	6 mm + 10 mm	PVB
0.89 mm	8 mm + 8 mm	Ionoplast
0.89 mm	10 mm + 6 mm	Ionoplast
0.89 mm	6 mm + 10 mm	Ionoplast
1.52 mm	8 mm + 8 mm	Ionoplast, PVB
1.52 mm	10 mm + 6 mm	Ionoplast, PVB
1.52 mm	6 mm + 10 mm	Ionoplast, PVB
2.28 mm	8 mm + 8 mm	Ionoplast, PVB
2.28 mm	10 mm + 6 mm	Ionoplast, PVB
2.28 mm	6 mm + 10 mm	Ionoplast, PVB

To obtain the relationship between the load duration, ambient temperature, and thickness of the interlayer compared to the maximum deflection (as a measure of flexural stiffness) and tension stress in the bottom glass plate, an analytical polynomial of three independent variables with unknown coefficients is used. Every functional independence of these three variables (load duration, ambient temperature, and thickness of interlayer) is defined by a second-order polynomial that generated a sixth-total-order polynomial:

$$P(x,y,z) = F_1 + F_2 \cdot x + F_3 \cdot y + F_4 \cdot x^2 + F_5 \cdot Fx \cdot y + F_6 \cdot y^2 + F_7 \cdot x^2 \cdot y + F_8 \cdot x \cdot y^2 + F_9 \cdot x^2 \cdot y^2 + F_{10} \cdot 0z + F_{11} \cdot 1x \cdot z + F_{12} \cdot y \cdot z + F_{13} \cdot x^2 \cdot xz + F_{14} \cdot x \cdot xy \cdot z + F_{15} \cdot y^2 \cdot yz + F_{16} \cdot 6x^2 \cdot 6y \cdot z + F_{17} \cdot 7x \cdot y^2 \cdot z + F_{18} \cdot 8x^2 \cdot 8y^2 \cdot 8z + F_{19} \cdot z^2 + F_{20} \cdot 0x \cdot z^2 + F_{21} \cdot 1y \cdot z^2 + F_{22} \cdot 2x^2 \cdot 2z^2 + F_{23} \cdot 3x \cdot y \cdot z^2 + F_{24} \cdot 4F^2 \cdot 4F^2 + F_{25} \cdot x^2 \cdot y \cdot z^2 + F_{26} \cdot x \cdot y^2 \cdot z^2 + F_{27} \cdot 7^2 \cdot y^2 \cdot z^2 \quad (1)$$

where x, y, and z are independent variables assigned to the physical values of the logarithm of load duration in seconds, ambient temperature in °C, and interlayer thickness in mm; the Fs are unknown coefficients.

The unknown coefficients are solved iteratively by MATLAB [39], and the nlinfit function is used. The coefficients are calculated using a nonlinear least-squares estimation [40]. The solution has converged successfully for all of the datasets.

For every fit, multiple fitting realizations were performed to avoid locking to the nearest target-function minimum instead of the global optimal point. The initial values of the coefficients were randomly chosen using a normal distribution, with the mean at 0 and a standard deviation of 1000. For every dataset, 100 realizations were performed. Every calculation converged up to four potential solutions. In the case of a single solution, it was chosen as the optimal one; in the case of multiple potential solutions, the optimal solution was chosen by the criterion of the lowest RMSE.

An example of the calculation is shown in Figure 8. In the top-left panel, a set of initial values are displayed, and in the top-right panel calculated values are displayed. The color of a point indicates a specific coefficient. All of the calculated solutions can be sorted into three unique solutions. The metrics displayed in the bottom panels are used as the criteria for the most successful fit. The solution with an RMSE of 0.1397 is chosen as the optimal one, as is displayed in the bottom-left panel. The R^2 metric in the bottom-right panel is in agreement with this choice, as the highest R^2 is with this solution.

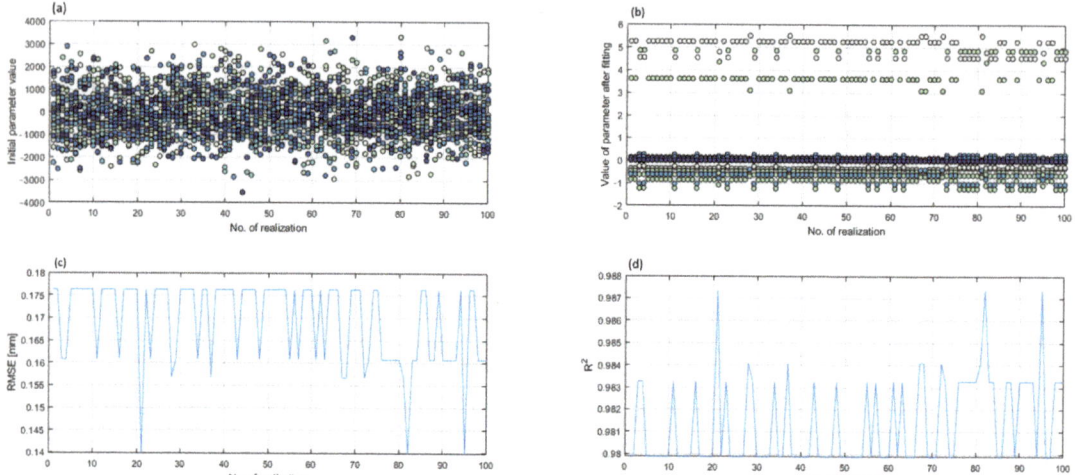

Figure 8. Panels representing an example of calculation: (**a**) initial parameter values; (**b**) calculated parameter values: (**c**) root-mean-square error (RMSE); and (**d**) determination coefficient.

The datasets used to obtain functions are divided into two classes of materials (PVB and ionoplast) and three classes of glass plate-thickness combinations. Every dataset consists of 48 gridded entries, four sets along the duration dimension, four along the temperature dimension, and three along the interlayer thickness dimension. The datasets are listed in Appendix A. The calculated coefficients for the displacement functions for the three glass plate-thickness combinations are listed in Appendix B.

The results presented in Figure 9 display a good fit between the provided data and the analytical model. The average error (RMSE) for the displacement function is less than 0.32 mm for PVB interlayers and 0.11 mm for ionoplast interlayers, and the RMSE for the stress function is under 0.93 MPa for PVB interlayers and less than 0.33 MPa for ionoplast interlayers. The coefficient of determination, R^2, is above 0.95 for both interlayers, which suggests a strong relationship between the data and the model. The almost-identical R^2 and the larger RMSE for PVB interlayers are due to the wider spread of results in PVB interlayers. The explanation for the discrepancies is numerical data errors in both numerical modeling and function fitting and a dismissed higher-order polynomial that were not taken into account due to the data dissipation.

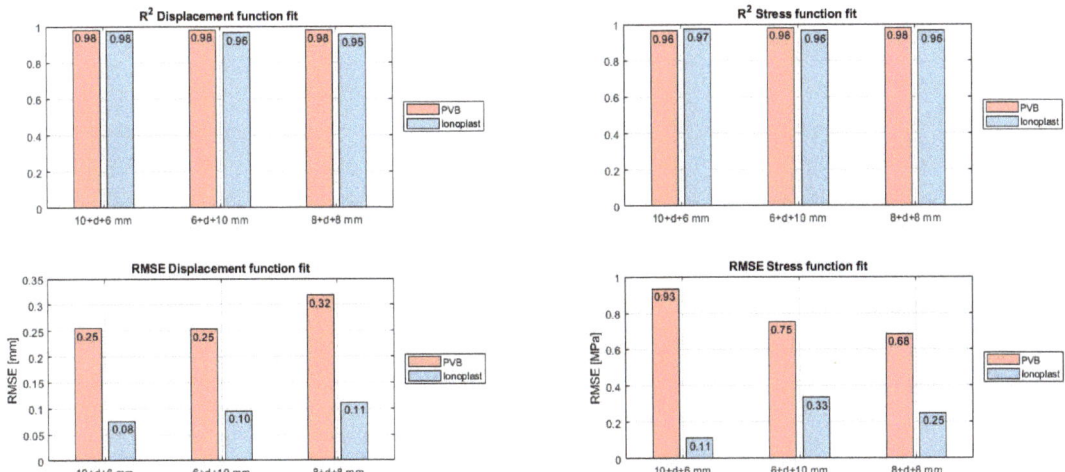

Figure 9. Fitting results of displacement (**left** panel) and stress function (**right** panel).

Finally, for all of the information, isosurfaces were created, showing the dependence of the deflection and stress on the specified parameters. The isosurfaces clearly show differences in the behavior of laminated glass panels along described independent variables at different plate-thickness dispositions and for different interlayers. Generally, ionoplast interlayers show higher flexural rigidity and better behavior when higher temperatures and load durations are involved. On the other hand, PVB interlayers have a lower capacity for these types of static loads. As we can see in Figure 10, glass panels with PVB interlayers show larger deflection (for long time loading and an increasing temperature this deflection is double) in comparison with ionoplast interlayers for the presented combinations (thickness of the glass panel, load duration, and temperature).

The total thickness of the glass in all of the presented models is 16 mm, but with different distributions of thickness in the glass panels: 6 mm + 10 mm; 8 mm + 8 mm; and 10 mm + 6 mm. We can notice that, for nonsymmetrical panels, the position of thinner and thicker glass plates (6 mm + 10 mm and 10 mm + 6 mm) does not provide the difference in the total deflection of the panels, and those values are slightly lower than the deflection from a symmetrical distribution of plates (8 mm + 8 mm) (Figures 10–12).

For the same laminated glass panels, appropriate isosurfaces for stresses are presented in Figures 13–15.

Analyzing stress in regard to thickness dispositions, we can notice that for nonsymmetrical glass plate dispositions, greater stress appears in cases where the thicker glass plate is at the bottom (6 mm + 10 mm) for both types of interlayers. The greatest difference is visible sfor a loading of one month at 30 °C. For example, in this case, the stresses in laminated glass with a disposition of 6 mm + 10 mm with PVB interlayers are 19.23% greater (Figures 13a and 15a) than those of laminated glass with a disposition of 10 mm + 6 mm. Additionally, for laminated glass panels with ionoplast interlayers, stresses are 7.05% greater than those for laminated glass with a disposition of 10 mm + 6 mm (Figures 13b and 15b). The symmetrical disposition (8 mm + 8 mm) shows values of stress between two presented limits (6 mm + 10 mm and 10 mm + 6 mm) (Figure 14).

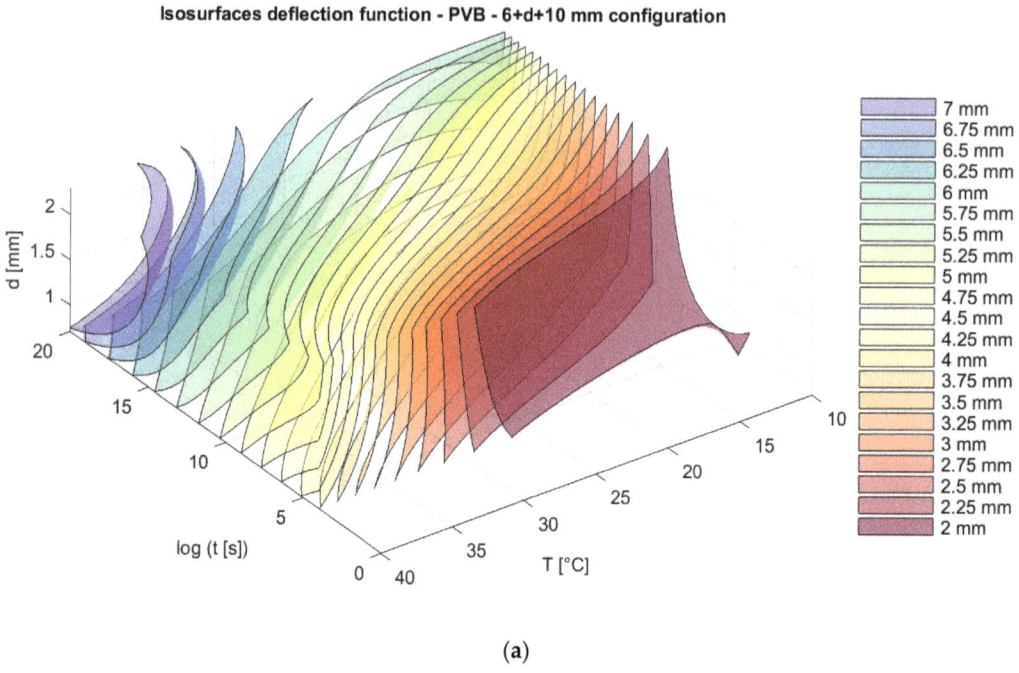

(a)

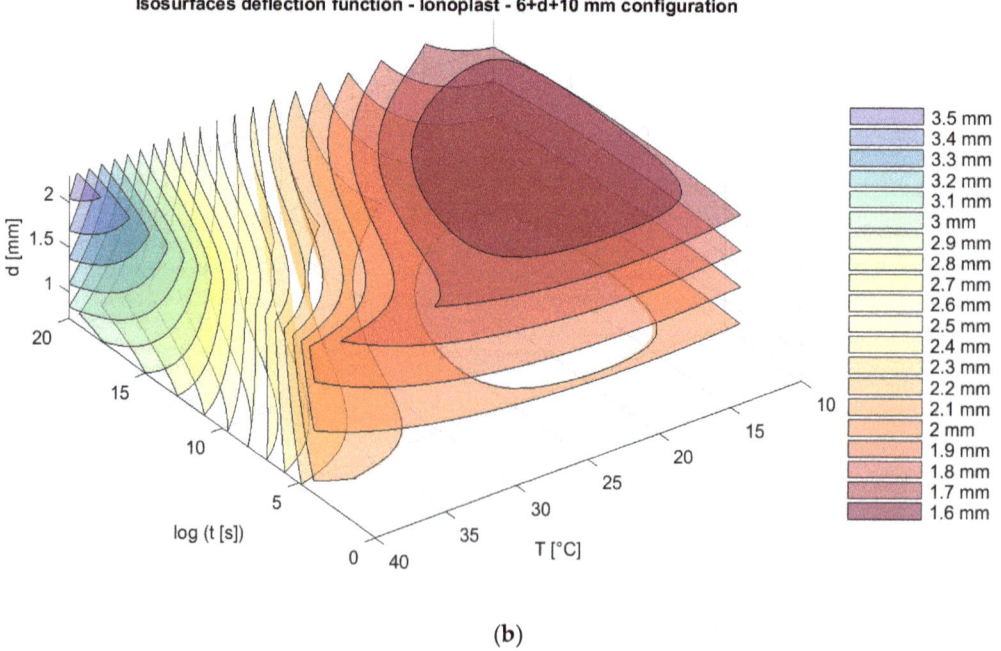

(b)

Figure 10. Isosurfaces for deflection in a model with an upper glass plate of 6 mm + a bottom glass plate of 10 mm with (**a**) PVB interlayers; (**b**) ionoplast interlayers.

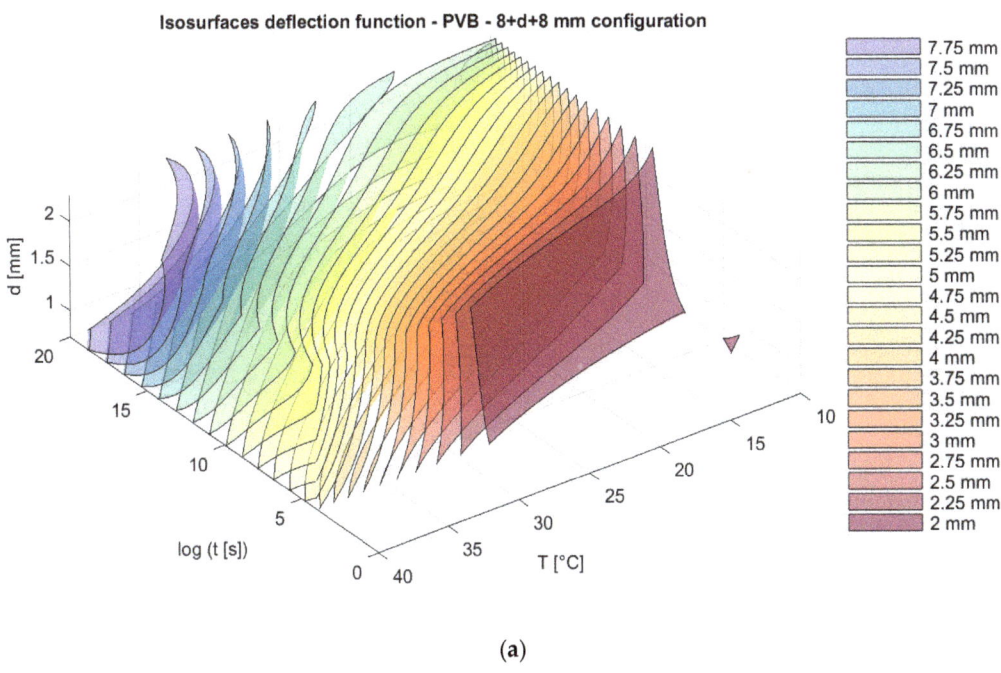

(a)

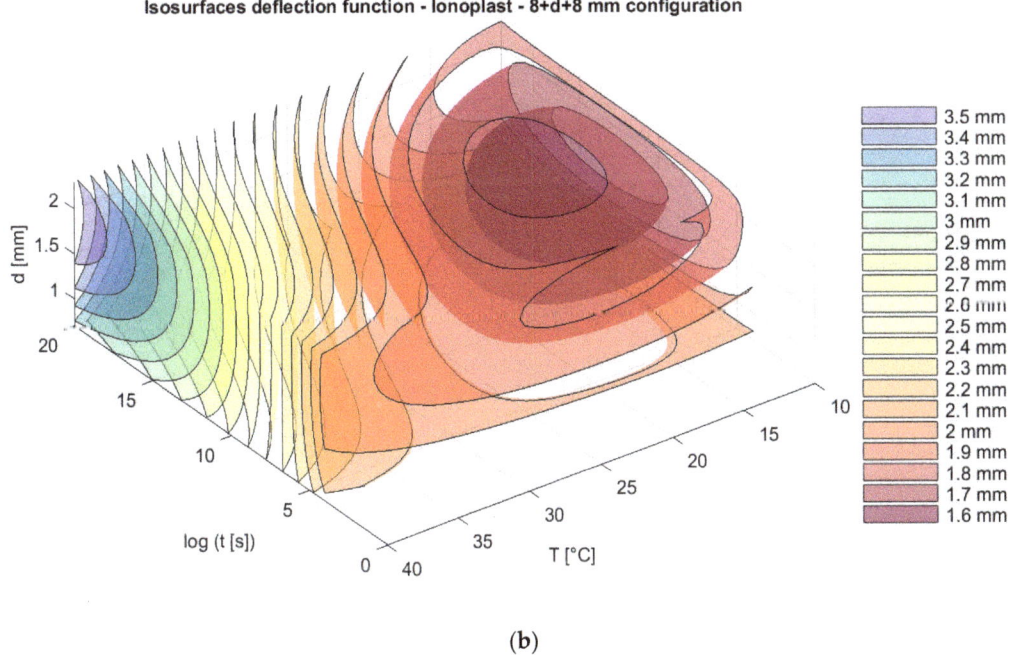

(b)

Figure 11. Isosurfaces for deflection in a model with an upper glass plate of 8 mm + a bottom glass plate of 8 mm with (**a**) PVB interlayers; (**b**) ionoplast interlayers.

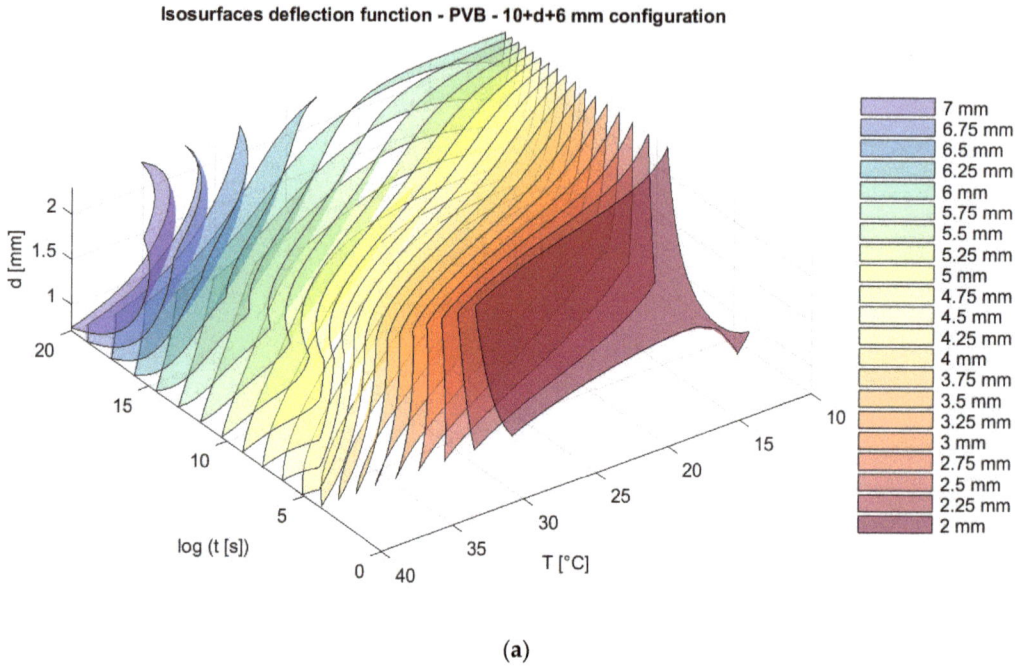

(a)

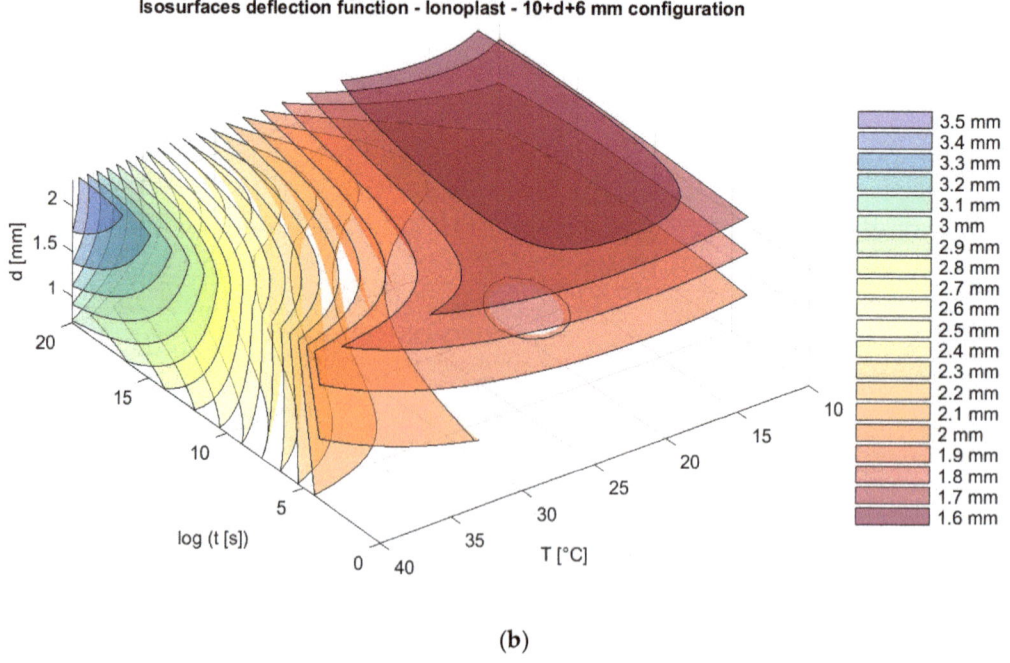

(b)

Figure 12. Isosurfaces for deflection in a model with an upper glass plate of 10 mm + a bottom glass plate of 6 mm with (**a**) PVB interlayers; (**b**) ionoplast interlayers.

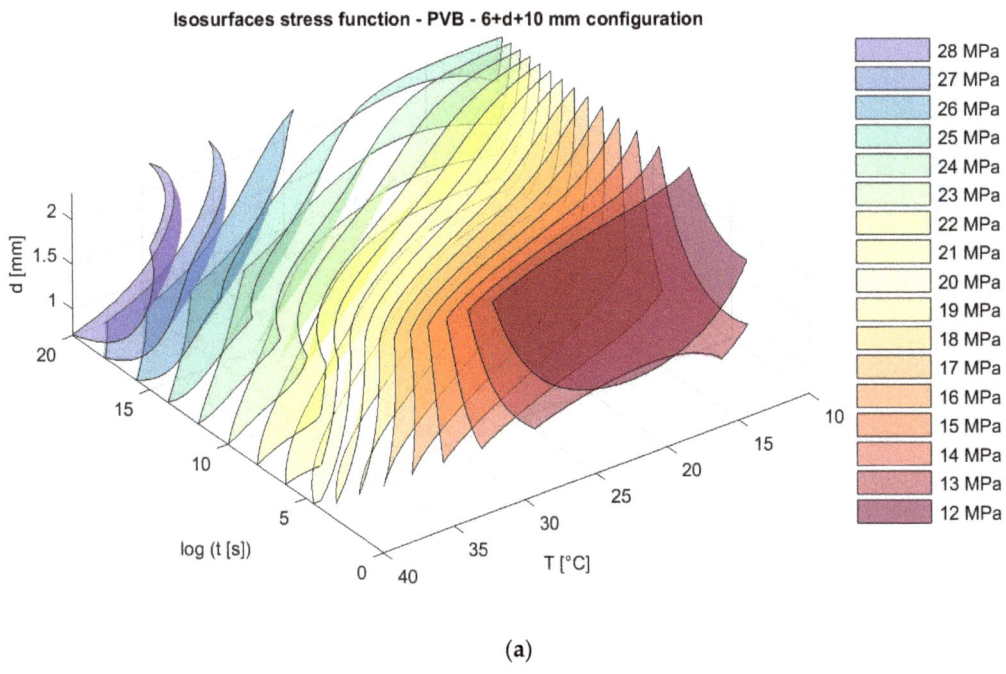

(a)

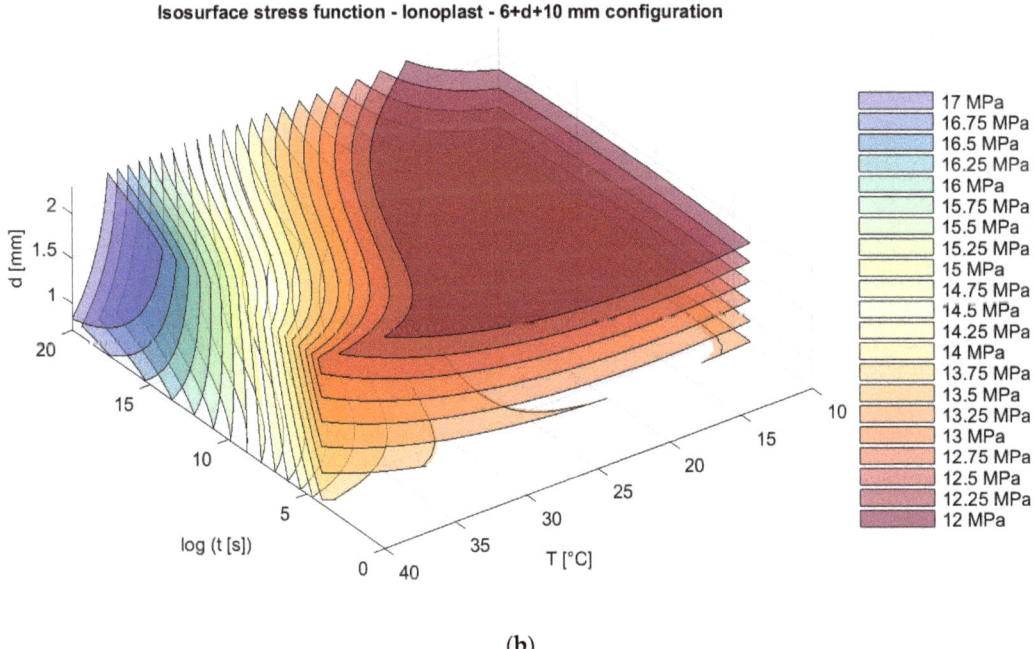

(b)

Figure 13. Isosurfaces for stress in a model with an upper glass plate of 6 mm + a bottom glass plate of 10 mm with (**a**) PVB interlayers; (**b**) ionoplast interlayers.

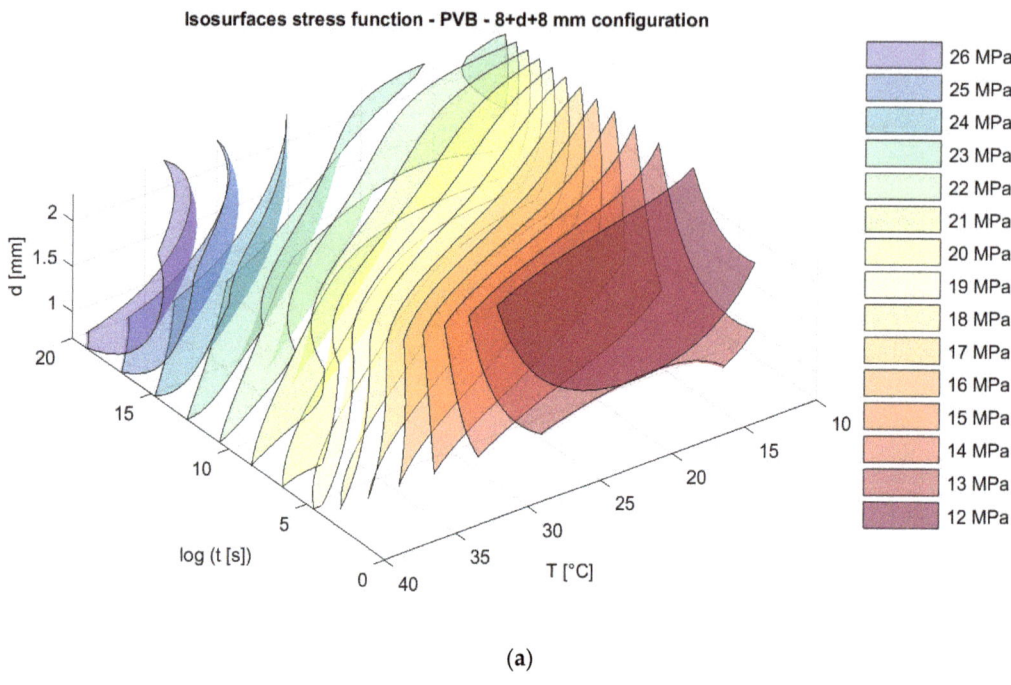

(a)

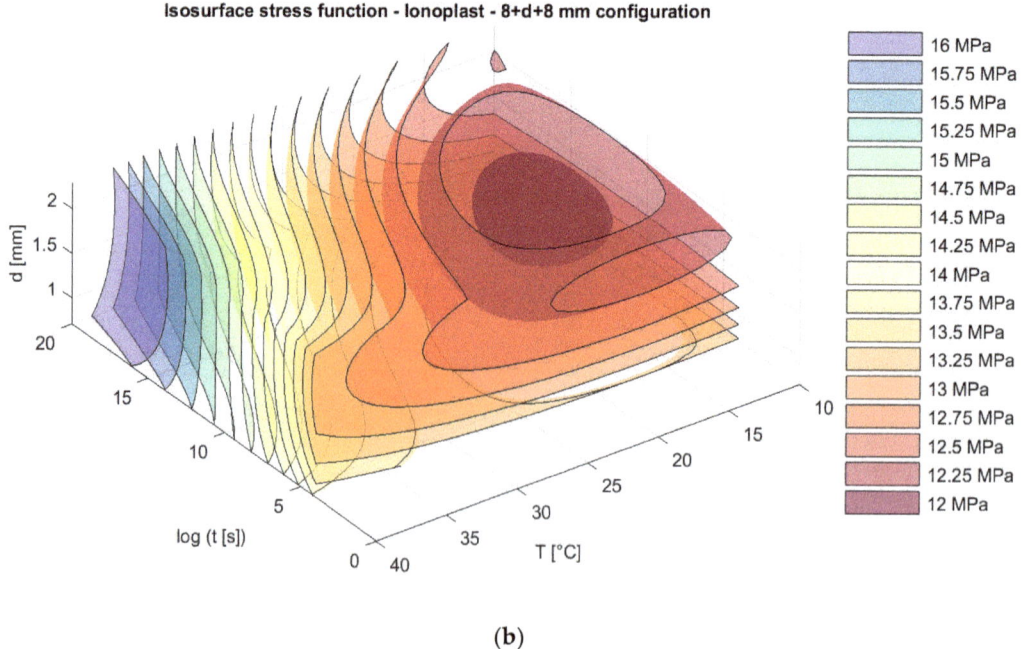

(b)

Figure 14. Isosurfaces for stress in a model with an upper glass plate of 8 mm + a bottom glass plate of 8 mm with (**a**) PVB interlayers; (**b**) ionoplast interlayers.

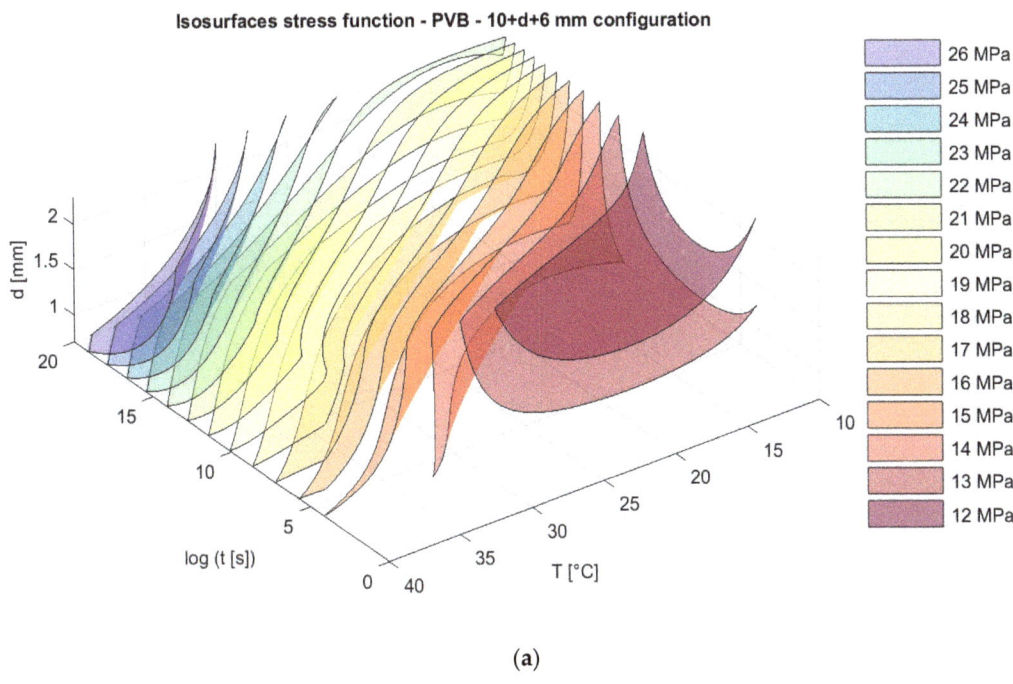

(a)

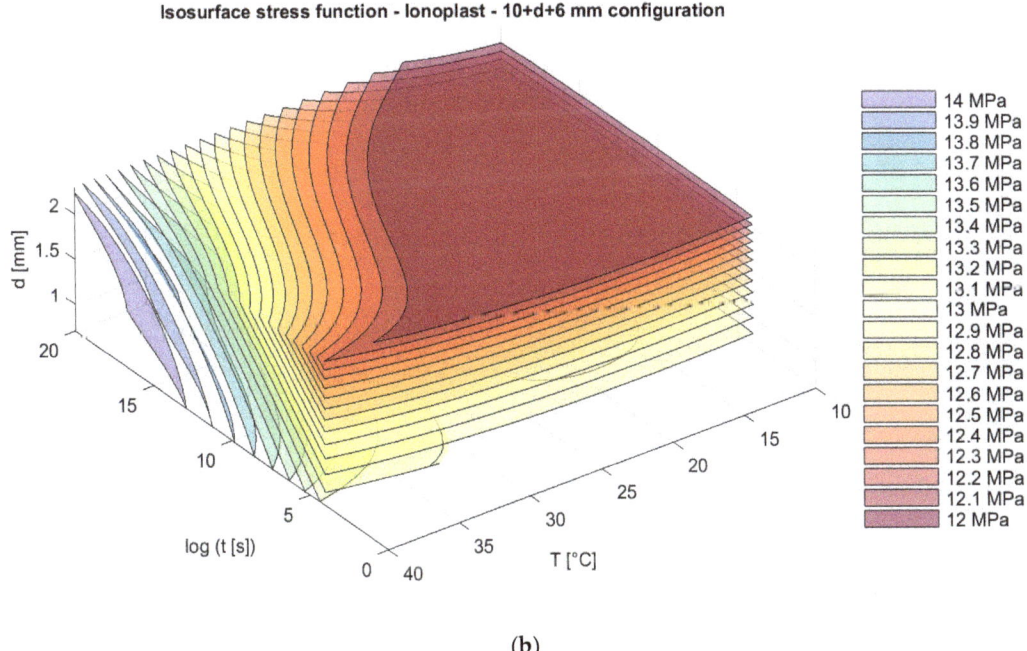

(b)

Figure 15. Isosurfaces for stress in a model with an upper glass plate of 10 mm + a bottom glass plate of 6 mm with (**a**) PVB interlayers; (**b**) ionoplast interlayers.

For temperatures up to 25 °C, an increase in interlayer thickness, for the ionoplast model, shows the positive effect of increasing the panel stiffness, resulting in lower deflection. For example, a glass panel with a disposition of 6 mm + 10 mm with an ionoplast interlayer thickness of 0.89 mm has 15.91% larger deflection compared to the same disposition glass panel with an ionoplast interlayer thickness of 2.28 mm. However, for temperatures above 35 °C, this effect is different, and for a long-term load the increase in interlayer thickness brings an 11.94% larger deflection of the panel when comparing an ionoplast interlayer thickness of 0.89 mm with one of 2.28 mm.

A different trend is observed in the model with PVB interlayers where it can be seen that, at lower temperatures (up to 25 °C) and shorter loadings, an increase in interlayer thickness does not provide any influence on the deflection. For longer loadings, an increase in the thickness of PVB interlayers shows an unfavorable effect, where deflection is approximately 13.62% higher when increasing the interlayer thickness from 0.76 mm to 2.28 mm when exposed to temperatures up to 25 °C. This difference between deflections decreases during higher temperatures (40 °C).

For long-term loading and increased temperatures, PVB interlayers show unfavorable effects regarding bearing capacity in comparison with stiffer interlayers, such as ionomers. These observations also match those of other experimental tests in the literature; see [41]. In graphs from Figures 10–15, we can see the benefits of ionoplast interlayers, but taking into consideration all other aspects, such as cost and availability, gives PVB interlayers the advantage, regardless of their slightly lower bearing capacity.

5. Conclusions

In this work, experimental and numerical analyses of laminated glass panels exposed to four-point bending (EN 1288-3:2000 [35]) are presented. First, the validation of the numerical model is performed, and then a model with different geometries is developed. Models composed of two glass plates with different thicknesses coupled with two types of interlayers are tested. The parametric study performed presents the influence of the temperature, load duration, type, and thickness of the interlayers, and the symmetric as well as nonsymmetric dispositions of the thicknesses of glass plates on the stresses and deflections of the laminated glass panels. To obtain the relationship between the load duration, temperature, and thickness of the interlayers compared to the maximum deflection (as a measure of flexural stiffness) and tension stress in the bottom glass plate, an analytical polynomial of three independent variables with unknown coefficients is defined. Every functional independence of these three variables (load duration, ambient temperature, and thickness of the interlayers) is defined by a second-order polynomial that generated a sixth-total-order polynomial. Isosurfaces are created, showing the dependence of the deflection and stress on the specified parameters, and clearly showing differences in the behavior of laminated glass panels for the same conditions with different interlayers.

The observed parameters have a significant influence on the bearing capacity of laminated glass panels, and they are of great importance in the design of glass structures. From the parametric study we can conclude that:

- Comparing deflections, laminated glass panels with ionoplast interlayers provide generally higher stiffness than the same model with PVB (Saflex DG41) interlayers does. This behavior is expected, because an ionoplast interlayer is an ionomer-based material that provides the highest level of structural performance [23], which contributes to increasing the stiffness of laminated glass and thus causes less deflection.
- Taking into account that the total thickness of glass in all of the models is 16 mm, we can see that, for nonsymmetrical panels, the position of thinner and thicker plates (6 mm + 10 mm and 10 mm + 6 mm) does not provide a difference in the total deflection of the panels, and these values are slightly lower than those from symmetrical dispositions of plates (8 mm + 8 mm). The dominant influence on the size of the deflection is the total thickness of the panels, which justifies the different approaches of calculating the effective thickness of the panels.
- When comparing the stresses on the bottom plate, we can see that due to the asymmetric dispositions of the plate, higher stresses occur in cases where the plate is thicker

at the bottom. This behavior is expected, because that part of the panel has a higher bending stiffness, withstanding higher load levels, resulting in higher stresses in that part. In the case of symmetrical dispositions (8 mm + 8 mm) of the panel, the expected stress values are between the previously mentioned two load limits for laminated glass with a panel disposition of 6 mm + 10 mm and 10 mm + 6 mm.

- For lower temperatures (up to 20 °C), an increase in the interlayer thickness, for the ionoplast model, shows the positive effect of increasing the panel stiffness, resulting in lower deflection. However, for temperatures above 35 °C, that effect disappears, and for long-term loading, the increase in the interlayer thickness brings a slightly higher deflection of the panel. A different trend is observed in the model with PVB (Saflex DG41, EASTMAN, Kingsport, TN, USA) interlayers, where it can be seen that at lower temperatures and shorter loadings, the increased thickness of the interlayers does not provide any influence on the deflection. The PVB (Saflex DG41) will be able to transfer a significant amount of shear force between the glass plates, and in this case the laminated panel has a bending stiffness of a similar magnitude to an equivalent monolithic panel of the same overall thickness.
- For higher temperatures and longer loadings, the increase in the interlayer thickness shows an unfavorable effect, where deflection is increased with an increased interlayer thickness. The same effect is visible for the load duration, where we observe greater deflections for longer load durations in both models, but this trend is more pronounced in models with PVB (Saflex DG41) interlayers. When this type of PVB is exposed to higher temperatures, its stiffness decreases, which is clearly manifested by the fall of Young's modulus, and it begins to behave in a similar manner to rubber, which results in an increased deflection.

Author Contributions: Conceptualization, M.G. and P.M.; methodology, M.G., G.G., V.D. and P.M.; software, V.D.; validation, M.G., G.G. and V.D.; investigation, M.G., G.G., V.D. and P.M.; resources, M.G., G.G., V.D. and P.M.; data curation, G.G.; writing—original draft preparation, M.G., G.G., V.D. and P.M.; writing—review and editing, M.G. and P.M.; visualization, G.G. and V.D.; supervision, M.G. and P.M.; project administration, M.G.; funding acquisition, M.G. and V.D. All authors have read and agreed to the published version of the manuscript.

Funding: This research was partially supported through project KK.01.1.1.02.0027, a project co-financed by the Croatian Government and the European Union through the European Regional Development Fund, the Competitiveness and Cohesion Operational Programme.

Institutional Review Board Statement: Not applicable.

Informed Consent Statement: Not applicable.

Data Availability Statement: Not applicable.

Acknowledgments: The authors gratefully acknowledge the support of the company DIALUX d.o.o. and Ivan Kuran from Split, Croatia for providing laminated glass samples for use in the experimental program and for giving technical assistance.

Conflicts of Interest: The authors declare no conflict of interest. The funders had no role in the design of the study; in the collection, analyses, or interpretation of the data; in the writing of the manuscript; or in the decision to publish the results.

Appendix A

The dataset with 48 gridded entries, consisting of four sets of load duration, four sets of temperature, and three sets of interlayer thickness, is presented in Tables A1–A3. The datasets are used to obtain functions, and they are divided into two classes of materials (PVB and ionoplast) and three classes of glass plate-thickness dispositions.

Table A1. Dataset for plate-thickness disposition 10 + d + 6 (mm).

Temp (°C)	Deflection 10 + 2.28 + 6 (mm)							
	PVB				IONOPLAST			
	1 min	24 h	1 month	10 years	1 min	24 h	1 month	10 years
10	1.642	1.960	4.185	5.737	1.645	1.652	1.657	1.660
25	1.673	4.255	5.586	6.141	1.660	1.690	1.739	1.820
30	1.920	5.580	6.048	6.753	1.684	1.770	2.232	2.081
40	5.200	6.320	6.996	7.421	1.865	2.964	3.329	3.467

Temp (°C)	Bottom panel stress 10 + 2.28 + 6 (MPa)							
	PVB				IONOPLAST			
	1 min	24 h	1 month	10 years	1 min	24 h	1 month	10 years
10	11.820	12.273	14.893	21.577	11.809	11.821	11.834	11.840
25	11.865	15.169	20.949	23.349	11.834	11.885	11.963	12.090
30	12.218	20.897	22.930	25.903	11.874	12.012	12.605	13.226
40	19.276	24.102	26.977	28.790	12.150	13.382	13.739	13.882

Temp (°C)	Deflection 10 + 1.52 + 6 (mm)							
	PVB				IONOPLAST			
	1 min	24 h	1 month	10 years	1 min	24 h	1 month	10 years
10	1.835	2.077	3.909	5.403	1.837	1.843	1.847	1.852
25	1.858	3.963	5.245	5.827	1.848	1.872	1.907	1.971
30	2.040	5.234	5.724	6.500	1.866	1.931	2.279	2.732
40	4.860	6.015	6.778	7.279	2.000	2.858	3.158	3.274

Temp (°C)	Bottom panel stress 10 + 1.52 + 6 (MPa)							
	PVB				IONOPLAST			
	1 min	24 h	1 month	10 years	1 min	24 h	1 month	10 years
10	12.614	12.910	14.676	19.998	12.605	12.612	12.621	12.624
25	12.641	14.731	19.339	21.891	12.620	12.652	12.703	12.783
30	12.886	19.266	21.429	24.810	12.646	12.732	13.159	13.615
40	17.619	22.710	26.000	28.180	12.822	13.733	14.000	14.113

Temp (°C)	Deflection 10 + 0.76 + 6 (mm)				Deflection 10 + 0.89 + 6 (mm)			
	PVB				IONOPLAST			
	1 min	24 h	1 month	10 years	1 min	24 h	1 month	10 years
10	2.064	2.210	3.457	4.736	1.978	1.981	1.984	1.988
25	2.077	3.447	4.561	5.157	1.984	2.000	2.023	2.065
30	2.181	4.525	5.031	5.907	1.996	2.038	2.266	2.576
40	4.177	5.323	6.241	6.922	2.083	2.664	2.881	2.967

Temp (°C)	Bottom panel stress 10 + 0.76 + 6 (MPa)				Bottom panel stress 10 + 0.89 + 6 (MPa)			
	PVB				IONOPLAST			
	1 min	24 h	1 month	10 years	1 min	24 h	1 month	10 years
10	13.508	13.630	14.740	16.885	13.133	13.138	13.140	13.145
25	13.516	14.725	16.105	18.756	13.142	13.161	13.185	13.232
30	13.605	15.964	18.195	22.075	13.157	13.203	13.481	13.813
40	15.320	19.515	23.544	26.557	13.256	13.898	14.103	14.182

Table A2. Dataset for plate-thickness disposition 6 + d + 10 (mm).

	Deflection 6 + 2.28 + 10 (mm)							
	PVB				IONOPLAST			
Temp (°C)	1 min	24 h	1 month	10 years	1 min	24 h	1 month	10 years
10	1.640	1.960	4.183	5.733	1.645	1.652	1.657	1.660
25	1.670	4.253	5.582	6.135	1.660	1.690	1.739	1.825
30	1.919	5.575	6.042	6.742	1.680	1.770	2.232	2.810
40	5.201	6.312	6.982	7.395	1.865	2.964	3.329	3.467
	Bottom panel stress 6 + 2.28 + 10 (MPa)							
	PVB				IONOPLAST			
Temp (°C)	1 min	24 h	1 month	10 years	1 min	24 h	1 month	10 years
10	11.440	12.784	19.698	24.349	11.444	11.490	11.522	11.559
25	11.619	19.910	23.896	25.557	11.536	11.710	11.940	12.301
30	12.631	23.876	25.274	27.401	11.674	12.083	13.761	15.617
40	22.746	26.094	28.146	29.511	12.461	16.096	17.201	17.612
	Deflection 6 + 1.52 + 10 (mm)							
	PVB				IONOPLAST			
Temp (°C)	1 min	24 h	1 month	10 years	1 min	24 h	1 month	10 years
10	1.835	2.077	3.907	5.402	1.837	1.843	1.847	1.852
25	1.858	3.962	5.242	5.824	1.848	1.872	1.907	1.971
30	2.042	5.231	5.720	6.493	1.866	1.931	2.279	2.732
40	4.855	6.010	6.768	7.260	2.001	2.857	3.158	3.274
	Bottom panel stress 6 + 1.52 + 10 (MPa)							
	PVB				IONOPLAST			
Temp (°C)	1 min	24 h	1 month	10 years	1 min	24 h	1 month	10 years
10	12.362	13.416	19.086	23.483	12.365	12.403	12.429	12.459
25	12.506	19.249	23.014	24.730	12.439	12.579	12.761	13.044
30	13.292	22.984	24.426	26.707	12.549	12.873	14.176	15.649
40	21.833	25.279	27.528	29.051	13.166	16.037	16.947	17.291
	Deflection 6 + 0.76 + 10 (mm)				Deflection 6 + 0.89 + 10 (mm)			
	PVB				IONOPLAST			
Temp (°C)	1 min	24 h	1 month	10 years	1 min	24 h	1 month	10 years
10	2.063	2.203	3.421	4.707	2.022	2.026	2.029	2.032
25	2.076	3.440	4.540	5.133	2.029	2.045	2.068	2.110
30	2.180	4.520	5.013	5.888	2.041	2.083	2.312	2.624
40	4.174	5.317	6.230	6.907	2.129	2.713	2.931	3.018
	Bottom panel stress 6 + 0.76 + 10 (MPa)				Bottom panel stress 6 + 0.89 + 10 (MPa)			
	PVB				IONOPLAST			
Temp (°C)	1 min	24 h	1 month	10 years	1 min	24 h	1 month	10 years
10	13.400	14.093	17.921	21.643	13.218	13.247	13.267	13.290
25	13.498	17.992	21.160	22.867	13.273	13.376	13.509	13.712
30	14.006	21.102	22.528	25.042	13.354	13.589	14.504	15.552
40	20.108	23.399	26.025	27.994	13.797	15.833	16.507	16.767

Table A3. Dataset for plate-thickness disposition 8 + d + 8 (mm).

Temp (°C)	Deflection 8 + 2.28 + 8 (mm)							
	PVB				IONOPLAST			
	1 min	24 h	1 month	10 years	1 min	24 h	1 month	10 years
10	1.614	1.971	4.223	6.060	1.815	1.821	1.825	1.831
25	1.647	4.621	5.851	6.597	1.827	1.852	1.892	1.963
30	1.918	6.266	6.460	7.800	1.846	1.918	2.308	2.823
40	5.790	7.224	7.834	8.691	1.996	2.966	3.314	3.450

Temp (°C)	Bottom panel stress 8 + 2.28 + 8 (MPa)							
	PVB				IONOPLAST			
	1 min	24 h	1 month	10 years	1 min	24 h	1 month	10 years
10	11.458	12.591	18.121	22.108	12.360	12.388	12.409	12.434
25	11.600	18.718	21.654	23.277	12.417	12.527	12.675	12.907
30	12.449	22.383	22.975	25.827	12.504	12.767	13.872	15.103
40	21.317	24.517	25.973	27.980	13.010	15.424	16.212	16.514

Temp (°C)	Deflection 8 + 1.52 + 8 (mm)							
	PVB				IONOPLAST			
	1 min	24 h	1 month	10 years	1 min	24 h	1 month	10 years
10	1.812	2.084	4.223	6.061	1.815	1.821	1.258	1.831
25	1.837	4.265	5.851	6.597	1.827	1.852	1.892	1.963
30	2.040	5.827	6.458	7.470	1.846	1.918	2.308	2.822
40	5.356	6.826	7.834	8.501	1.996	2.966	3.314	3.449

Temp (°C)	Bottom panel stress 8 + 1.52 + 8 (MPa)							
	PVB				IONOPLAST			
	1 min	24 h	1 month	10 years	1 min	24 h	1 month	10 years
10	12.361	13.225	18.121	22.108	12.360	12.390	12.409	12.435
25	12.471	18.217	21.654	23.227	12.417	12.528	12.675	12.907
30	13.111	21.603	22.975	25.171	12.504	12.767	13.872	15.103
40	20.580	23.776	25.973	27.534	13.010	15.424	16.212	16.514

Temp (°C)	Deflection 8 + 0.76 + 8 (mm)				Deflection 8 + 0.89 + 8 (mm)			
	PVB				IONOPLAST			
	1 min	24 h	1 month	10 years	1 min	24 h	1 month	10 years
10	2.049	2.211	3.644	5.197	2.007	2.011	2.014	2.019
25	2.063	3.460	4.979	5.721	2.015	2.032	2.058	2.105
30	2.018	4.939	5.564	6.681	2.028	2.075	2.333	2.688
40	4.514	5.937	7.119	8.027	2.125	2.789	3.042	3.142

Temp (°C)	Bottom panel stress 8 + 0.76 + 8 (MPa)				Bottom panel stress 8 + 0.89 + 8 (MPa)			
	PVB				IONOPLAST			
	1 min	24 h	1 month	10 years	1 min	24 h	1 month	10 years
10	13.385	13.931	17.225	20.514	13.200	13.222	13.236	13.258
25	13.456	17.224	20.060	21.615	13.241	13.321	13.425	13.590
30	13.850	19.971	21.285	23.633	13.303	13.490	14.244	15.131
40	19.071	22.057	24.554	26.474	13.659	15.365	15.926	16.147

Appendix B

The calculated coefficients for displacement and stress functions for three glass plate-thickness dispositions are presented in Tables A4 and A5.

Table A4. Calculated coefficients for displacement functions (1) for three plate-thickness dispositions.

Coefficients for Displacement Function

Coef.	10 + d + 6 mm PVB	6 + d + 10 mm PVB	8 + d + 8 mm PVB	10 + d + 6 mm Ionoplast	6 + d + 10 mm Ionoplast	8 + d + 8 mm Ionoplast
F_1	2.791×10^0	2.871×10^0	2.910×10^0	2.263×10^0	1.992×10^0	-2.809×10^{-1}
F_2	3.290×10^{-2}	1.531×10^{-2}	1.750×10^{-1}	-5.358×10^{-2}	6.979×10^{-2}	6.894×10^{-1}
F_3	8.810×10^{-3}	2.529×10^{-3}	1.416×10^{-2}	-1.987×10^{-2}	2.648×10^{-2}	1.708×10^{-1}
F_4	-2.880×10^{-3}	-2.442×10^{-3}	-1.257×10^{-2}	4.104×10^{-3}	-2.826×10^{-3}	-2.654×10^{-2}
F_5	-2.783×10^{-2}	-2.660×10^{-2}	-4.955×10^{-2}	5.285×10^{-3}	-8.877×10^{-3}	-4.392×10^{-2}
F_6	-1.292×10^{-3}	-1.168×10^{-3}	-1.699×10^{-3}	2.399×10^{-4}	-6.416×10^{-4}	-2.636×10^{-3}
F_7	1.836×10^{-3}	1.787×10^{-3}	3.079×10^{-3}	-4.501×10^{-4}	3.457×10^{-4}	1.675×10^{-3}
F_8	1.030×10^{-3}	1.007×10^{-3}	1.553×10^{-3}	-5.441×10^{-5}	2.151×10^{-4}	6.741×10^{-4}
F_9	-5.742×10^{-5}	-5.634×10^{-5}	-8.522×10^{-5}	7.458×10^{-6}	-7.622×10^{-6}	-2.530×10^{-5}
F_{10}	5.708×10^0	5.599×10^0	7.015×10^0	-1.170×10^0	-5.342×10^{-1}	3.249×10^0
F_{11}	-1.458×10^0	-1.436×10^0	-1.959×10^0	3.391×10^{-1}	9.983×10^{-2}	-9.423×10^{-1}
F_{12}	-7.799×10^{-1}	-7.695×10^{-1}	-9.473×10^{-1}	1.290×10^{-1}	3.911×10^{-2}	-1.964×10^{-1}
F_{13}	7.167×10^{-2}	7.108×10^{-2}	1.003×10^{-1}	-1.755×10^{-2}	-4.043×10^{-3}	3.575×10^{-2}
F_{14}	1.785×10^{-1}	1.767×10^{-1}	2.359×10^{-1}	-4.042×10^{-2}	-1.288×10^{-2}	4.278×10^{-2}
F_{15}	1.828×10^{-2}	1.805×10^{-2}	2.206×10^{-2}	-2.657×10^{-3}	-9.483×10^{-4}	2.119×10^{-3}
F_{16}	-3.968×10^{-3}	-3.928×10^{-3}	-5.153×10^{-3}	8.385×10^{-4}	3.154×10^{-4}	-3.478×10^{-4}
F_{17}	-8.110×10^{-3}	-8.036×10^{-3}	-1.095×10^{-2}	2.063×10^{-3}	5.083×10^{-4}	-1.595×10^{-3}
F_{18}	1.814×10^{-4}	1.796×10^{-4}	2.395×10^{-4}	-4.088×10^{-5}	-1.143×10^{-5}	1.449×10^{-5}
F_{19}	-1.353×10^0	-1.321×10^0	-1.760×10^0	3.169×10^{-1}	1.684×10^{-2}	-1.157×10^0
F_{20}	3.504×10^{-1}	3.443×10^{-1}	5.087×10^{-1}	-1.093×10^{-1}	-5.482×10^{-3}	3.262×10^{-1}
F_{21}	1.748×10^{-1}	1.714×10^{-1}	2.248×10^{-1}	-4.197×10^{-2}	-2.992×10^{-3}	7.059×10^{-2}

Table A4. Cont.

Coefficients for Displacement Function

Coef.	10 + d + 6 mm PVB	6 + d + 10 mm PVB	8 + d + 8 mm PVB	10 + d + 6 mm Ionoplast	6 + d + 10 mm Ionoplast	8 + d + 8 mm Ionoplast
F_{22}	-1.779×10^{-2}	-1.763×10^{-2}	-2.709×10^{-2}	6.027×10^{-3}	1.461×10^{-4}	-1.245×10^{-2}
F_{23}	-4.278×10^{-2}	-4.220×10^{-2}	-6.037×10^{-2}	1.270×10^{-2}	7.323×10^{-4}	-1.615×10^{-2}
F_{24}	-4.177×10^{-3}	-4.100×10^{-3}	-5.248×10^{-3}	8.110×10^{-4}	7.145×10^{-5}	-8.424×10^{-4}
F_{25}	9.932×10^{-4}	9.803×10^{-4}	1.350×10^{-3}	-2.447×10^{-4}	-1.785×10^{-5}	1.653×10^{-4}
F_{26}	2.078×10^{-3}	2.055×10^{-3}	2.963×10^{-3}	-6.981×10^{-4}	-2.049×10^{-5}	6.098×10^{-4}
F_{27}	-4.897×10^{-5}	-4.843×10^{-5}	-6.678×10^{-5}	1.328×10^{-5}	4.615×10^{-7}	-6.677×10^{-6}

Table A5. The calculated coefficients for stress function (1) for three plate-thickness dispositions.

Coefficients for Stress Function

Coef.	10 + d + 6 mm PVB	6 + d + 10 mm PVB	8 + d + 8 mm PVB	10 + d + 6 mm Ionoplast	6 + d + 10 mm Ionoplast	8 + d + 8 mm Ionoplast
F_1	2.791×10^{0}	2.871×10^{0}	2.910×10^{0}	2.263×10^{0}	1.992×10^{0}	-2.809×10^{-1}
F_2	3.290×10^{-2}	1.531×10^{-2}	1.750×10^{-1}	-5.358×10^{-2}	6.979×10^{-2}	6.894×10^{-1}
F_3	8.810×10^{-3}	2.529×10^{-3}	1.416×10^{-2}	-1.987×10^{-2}	2.648×10^{-2}	1.708×10^{-1}
F_4	-2.880×10^{-3}	-2.442×10^{-3}	-1.257×10^{-2}	4.104×10^{-3}	-2.826×10^{-3}	-2.654×10^{-2}
F_5	-2.783×10^{-2}	-2.660×10^{-2}	-4.955×10^{-2}	5.285×10^{-3}	-8.877×10^{-3}	-4.392×10^{-2}
F_6	-1.292×10^{-3}	-1.168×10^{-3}	-1.699×10^{-3}	2.399×10^{-4}	-6.416×10^{-4}	-2.636×10^{-3}
F_7	1.836×10^{-3}	1.787×10^{-3}	3.079×10^{-3}	-4.501×10^{-4}	3.457×10^{-4}	1.675×10^{-3}
F_8	1.030×10^{-3}	1.007×10^{-3}	1.553×10^{-3}	-5.441×10^{-5}	2.151×10^{-4}	6.741×10^{-4}
F_9	-5.742×10^{-5}	-5.634×10^{-5}	-8.522×10^{-5}	7.458×10^{-6}	-7.622×10^{-6}	-2.530×10^{-5}
F_{10}	5.708×10^{0}	5.599×10^{0}	7.015×10^{0}	-1.170×10^{0}	-5.342×10^{-1}	3.249×10^{0}

Table A5. Cont.

Coefficients for Stress Function

Coef.	10 + d + 6 mm PVB	6 + d + 10 mm PVB	8 + d + 8 mm PVB	10 + d + 6 mm Ionoplast	6 + d + 10 mm Ionoplast	8 + d + 8 mm Ionoplast
F_{11}	-1.458×10^{0}	-1.436×10^{0}	-1.959×10^{0}	3.391×10^{-1}	9.983×10^{-2}	-9.423×10^{-1}
F_{12}	-7.799×10^{-1}	-7.695×10^{-1}	-9.473×10^{-1}	1.290×10^{-1}	3.911×10^{-2}	-1.964×10^{-1}
F_{13}	7.167×10^{-2}	7.108×10^{-2}	1.003×10^{-1}	-1.755×10^{-2}	-4.043×10^{-3}	3.575×10^{-2}
F_{14}	1.785×10^{-1}	1.767×10^{-1}	2.359×10^{-1}	-4.042×10^{-2}	-1.288×10^{-2}	4.278×10^{-2}
F_{15}	1.828×10^{-2}	1.805×10^{-2}	2.206×10^{-2}	-2.657×10^{-3}	-9.483×10^{-4}	2.119×10^{-3}
F_{16}	-3.968×10^{-3}	-3.928×10^{-3}	-5.153×10^{-3}	8.385×10^{-4}	3.154×10^{-4}	-3.478×10^{-4}
F_{17}	-8.110×10^{-3}	-8.036×10^{-3}	-1.095×10^{-2}	2.063×10^{-3}	5.083×10^{-4}	-1.595×10^{-3}
F_{18}	1.814×10^{-4}	1.796×10^{-4}	2.395×10^{-4}	-4.088×10^{-5}	-1.143×10^{-5}	1.449×10^{-5}
F_{19}	-1.353×10^{0}	-1.321×10^{0}	-1.760×10^{0}	3.169×10^{-1}	1.684×10^{-2}	-1.157×10^{0}
F_{20}	3.504×10^{-1}	3.443×10^{-1}	5.087×10^{-1}	-1.093×10^{-1}	-5.482×10^{-3}	3.262×10^{-1}
F_{21}	1.748×10^{-1}	1.714×10^{-1}	2.248×10^{-1}	-4.197×10^{-2}	-2.992×10^{-3}	7.059×10^{-2}
F_{22}	-1.779×10^{-2}	-1.763×10^{-2}	-2.709×10^{-2}	6.027×10^{-3}	1.461×10^{-4}	-1.245×10^{-2}
F_{23}	-4.278×10^{-2}	-4.220×10^{-2}	-6.037×10^{-2}	1.270×10^{-2}	7.323×10^{-4}	-1.615×10^{-2}
F_{24}	-4.177×10^{-3}	-4.100×10^{-3}	-5.248×10^{-3}	8.110×10^{-4}	7.145×10^{-4}	-8.424×10^{-4}
F_{25}	9.932×10^{-4}	9.803×10^{-4}	1.350×10^{-3}	-2.447×10^{-4}	-1.785×10^{-5}	1.653×10^{-4}
F_{26}	2.078×10^{-3}	2.055×10^{-3}	2.963×10^{-3}	-6.981×10^{-4}	-2.049×10^{-5}	6.098×10^{-4}
F_{27}	-4.897×10^{-5}	-4.343×10^{-5}	-6.678×10^{-5}	1.328×10^{-5}	4.615×10^{-7}	-6.677×10^{-6}

References

1. *FprEN 16612*; Glass in Building—Determination of the Lateral Load Resistance of Glass Panes by Calculation. EUROPEAN STANDARDS s.r.o.: Pilsen, Czech Republic, 2019.
2. Grozdanić, G.; Galić, M.; Marović, P. Some Aspects of the Analyses of Glass Structures Exposed to Impact Load. *Couple. Syst. Mech.* **2021**, *10*, 475–490. [CrossRef]
3. Schmidt, J.; Zemanová, A.; Zeman, J.; Šejnoha, M. Phase-Field Fracture Modelling of Thin Monolithic and Laminated Glass Plates under Quasi-Static Bending. *Materials* **2020**, *13*, 5153. [CrossRef]
4. Zhang, X.; Liu, H.; Maharaj, C.; Zheng, M.; Mohagheghian, I.; Zhang, G.; Yan, Y.; Dear, J.P. Impact Response of Laminated Glass with Varying Interlayer Materials. *Int. J. Impact Eng.* **2020**, *139*, 103505:1–103505:15. [CrossRef]
5. Timmel, M.; Kolling, S.; Osterrieder, P.; Du Bois, P.A. A Finite Element Model for Impact Simulation with Laminated Glass. *Int. J. Impact Eng.* **2007**, *34*, 1465–1478. [CrossRef]
6. Serafinavicius, T.; Kvedaras, A.K.; Sauciuvenas, G. Bending Behavior of Structural Glass Laminated with Different Interlayers. *Mech. Compos. Mater.* **2013**, *49*, 437–446. [CrossRef]
7. Pelfrene, J.; Kuntsche, J.; Van Dam, S.; Van Paepegem, W.; Schneider, J. Critical Assessment of the Post-breakage Performance of Blast Loaded Laminated Glazing: Experiments and Simulations. *Int. J. Impact Eng.* **2016**, *88*, 61–71. [CrossRef]
8. Pankhardt, K.; Balázs, G.L. Temperature Dependent Load Bearing Capacity of Laminated Glass Panes. *Period. Polytech. Civil Engng.* **2010**, *54*, 11–22. [CrossRef]
9. Galuppi, L.; Royer-Carfagni, G.F. Effective Thickness of Laminated Glass Beams: New Expression via a Variational Approach. *Eng. Struct.* **2012**, *38*, 53–67. [CrossRef]
10. Van Duser, A.; Jagota, A.; Bennison, S.J. Analysis of Glass/Polyvinyl Butyral Laminates Subjected to Uniform Pressure. *ASCE J. Eng. Mech.* **1999**, *125*, 435–442. [CrossRef]
11. Louter, C.; Belis, J.; Veer, F.; Lebet, J.-P. Durability of SG-Laminated Reinforced Glass Beams: Effects of Temperature, Thermal Cycling, Humidity and Load-duration. *Constr. Build. Mater.* **2012**, *27*, 280–292. [CrossRef]
12. Schwarzl, F.R. *Polymer-Mechanik: Struktur und Mechanisches Verhalten von Polymeren*; Springer: Berlin/Heidelberg, Germany, 1990; ISBN 978-3-642-64858-8.
13. Brinson, H.F.; Brinson, L.C. *Polymer Engineering Science and Viscoelasticity: An Introduction*, 2nd ed.; Springer: New York, NY, USA, 2015; ISBN 978-1-4899-7484-6.
14. Hána, T.; Janda, T.; Schmidt, J.; Zemanova, A.; Šejnoha, M.; Eliášová, M.; Vokáč, M. Experimental and Numerical Study of Viscoelastic Properties of Polymeric Interlayers Used for Laminated Glass: Determination of Material Parameters. *Materials* **2019**, *12*, 2241. [CrossRef]
15. Hána, T.; Eliášová, M.; Machalická, K.; Vokáč, M. Determination of PVB Interlayer's Shear Modulus and its Effect on Normal Stress Distribution in Laminated Glass Panels. *Mater. Sci. Engng.* **2017**, *251*, 012076:1–012076:8. [CrossRef]
16. Biolzi, L.; Cattaneo, S.; Orlando, M.; Piscitelli, L.R.; Spinelli, P. Constitutive Relationships of Different Interlayer Materials for Laminated Glass. *Compos. Struct.* **2020**, *244*, 112221:1–112221:16. [CrossRef]
17. Iwasaki, R.; Sato, C.; Lataillade, J.L.; Viot, P. Experimental Study on the Interface Fracture Toughness of PVB (Polyvinyl Butyral)/Glass at High Strain Rates. *Int. J. Crashworthiness* **2007**, *12*, 293–298. [CrossRef]
18. Molnár, G.; Vigh, L.G.; Stocker, G.; Dunai, L. Finite Element Analysis of Laminated Structural Glass Plates with Polyvinyl Butyral (PVB) Interlayer. *Period. Polytech. Civil Engng.* **2012**, *56*, 35–42. [CrossRef]
19. *EASTMAN-Material Properties of PVB Interlayers Used in Saflex DG41, Product Sheets*; Saflex DG: Springfield, MA, USA, 2015.
20. SentryGlas® Ionoplast Interlayer-Elastic Properties (SG5000). Available online: https://www.trosifol.com/fileadmin/user_upload/technical_information/downloads/sentryglas/150129_Kuraray_TM_Datenblatt_SG.pdf (accessed on 29 January 2022).
21. Hooper, P.A.; Blackman, B.R.K.; Dear, J.P. The Mechanical Behaviour of Poly (Vinyl Bbutyral) at Different Strain Magnitudes and Strain Rates. *J. Mater. Sci.* **2012**, *47*, 3564–3576. [CrossRef]
22. Zhang, H.; Hao, H.; Shi, Y.; Cui, J. The Mechanical Properties of Polyvinyl Butyral (PVB) at High Strain Rates. *Constr. Build. Mater.* **2015**, *93*, 404–415. [CrossRef]
23. Centelles, X.; Martin, M.; Solé, A.; Castro, J.R.; Cabeza, L.F. Tensile Test on Interlayer Materials for Laminated Glass under Diverse Ageing Conditions and Strain Rates. *Constr. Build. Mater.* **2020**, *243*, 118230. [CrossRef]
24. Novotný, M.; Poot, B. Influence of temperature on laminated glass performances assembled with various interlayers. In Proceedings of the Conference on Architectural and Structural Applications of Glass (Challeging Glass 5), Ghent, Belgium, 16–17 June 2016.
25. Aenlle-Lopez, M.; Noriega, A.; Pelayo, F. Mechanical Characterization of Polyvinil Butyral from Static and Modal Tests on Laminated Glass Beams. *Compos. Part B Eng.* **2019**, *169*, 9–18. [CrossRef]
26. Liene, S.; Kinsella, D.; Kozłowski, M. Influence of EVA, PVB and Ionoplast Interlayers on the Structural Behaviour and Fracture Pattern of Laminated Glass. *Int. J. Struct. Glas. Adv. Mater. Res.* **2019**, *3*, 62–78. [CrossRef]
27. Schuster, M.; Kraus, M.; Schneider, J.; Siebert, G. Investigations on the Thermorheologically Complex Material Behaviour of the laminated Safety Glass Interlayer Ethylene-Vinyl-Acetate. *Glass Struct. Eng.* **2018**, *3*, 373–388. [CrossRef]
28. Kraus, M.A.; Schuster, M.; Kuntsche, J.; Siebert, G.; Schneider, J. Parameter Identification Methods for Visco- and Hyperelastic Material Models. *Glass Struct. Eng.* **2017**, *2*, 147–167. [CrossRef]

29. Hána, T.; Vokáč, M.; Eliášová, M.; Machalická, K.V. Experimental Investigation of Temperature and Loading Rate Effects on the Initial Shear Stiffness of Polymeric Interlayers. *Eng. Struct.* **2020**, *223*, 110728:1–110728:16. [CrossRef]
30. Biolzi, L.; Cagnacci, E.; Orlando, M.; Piscitelli, L.; Rosati, G. Long Term Response of Glass-PVB double-lap joints. *Compos. Part B Eng.* **2014**, 41–49. [CrossRef]
31. Andreozzi, L.; Briccoli Bati, S.; Fagone, M.; Ranocchiai, G.; Zulli, F. Dynamic Torsion Tests to Characterize the Termo-viscoelastic Properties of Polymeric Interlayers for Laminated Glass. *Constr. Build. Mater.* **2014**, *65*, 1–13. [CrossRef]
32. Botz, M.; Kraus, M.; Siebert, G. Experimental determination of the shear modulus of polymeric interlayers used in laminated glass. In Proceedings of the Glass Con. Global, Chicago, IL, USA, 5–7 September 2018.
33. Chen, S.; Lu, Y.; Zhang, Y.; Shao, X. Experimental and Analytical Study on Uniaxial Tensile Property of Ionomer Interlayer at Different Temperatures and Strain Rates. *Constr. Build. Mater.* **2020**, *262*, 120058:1–120058:18. [CrossRef]
34. Castori, G.; Speranzini, E. Structural Analysis of Failure Behavior of Laminated Glass. *Compos. Part B Eng.* **2017**, *125*, 89–99. [CrossRef]
35. EN 1288-3:2002; Glass in Building-Determination of the Bending Strength of Glass-Part 3: Test with Specimen Supported at Two Points (Four-Point Bending). European Committee for Standardization, CEN: Brussels, Belgium, 2000.
36. Asik, M.Z. Laminated Glass Plates: Revealing of Nonlinear Behavior. *Comput. Struct.* **2003**, *81*, 2659–2671. [CrossRef]
37. Gao, W.; Zang, M. The Simulation of Laminated Glass Beam Impact Problem by Developing Fracture Model of Spherical DEM. *Eng. Anal. Bound. Elem.* **2014**, *42*, 2–7. [CrossRef]
38. *Engineering Simulation Software ANSYS*, Release 16.2; Ansys Inc.: Canonsburg, DC, USA, 2015.
39. *Mathematical Computing Software MATLAB*, version 2021a; The MathWorks Inc.: Natick, MA, USA, 2021.
40. Seber, G.A.F.; Wild, C.J. *Nonlinear Regression*; John Wiley & Sons: Hoboken, NJ, USA, 2003; ISBN 0-471-47135-6.
41. Serafinavičius, T.; Lebet, J.-P.; Louter, C.; Lenkimas, T.; Kuranovas, A. Long-term Laminated Glass Four Point Bending Test with PVB, EVA and SG Interlayers at Different Temperatures. *Procedia Eng.* **2013**, *57*, 996–1004. [CrossRef]

Article

A Molten Salt Electrochemical Process for the Preparation of Cost-Effective p-Block (Coating) Materials

Prabhat Kumar Tripathy [1,*] and Kunal Mondal [2]

1 Pyrochemistry & Molten Salt Systems Department, Fuel Cycle Science and Technology Division, Nuclear Science and Technology Directorate, Idaho National Laboratory, Idaho Falls, ID 83415, USA
2 Materials Science and Engineering Department, Energy and Environment Science and Technology Directorate, Idaho National Laboratory, Idaho Falls, ID 83415, USA; kunal.mondal@inl.gov
* Correspondence: prabhat.tripathy@inl.gov

Abstract: Solar energy applications rely heavily on p-block elements and transition metals. Silicon is, by far, the most commonly used material in photovoltaic cells and accounts for about 85% of modules sold presently. Of late, thin film photovoltaic cells have gained momentum because of their higher efficiencies. Most of these thin film devices are made out of just five elements, namely, cadmium, tellurium, selenium, indium, gallium and copper. The present manuscript describes an elegant and inexpensive molten salt-based electrolytic process for fabricating a tellurium-coated metallic substrate. A three-electrode set up was employed to coat iridium with tellurium from a molten bath containing lithium chloride, lithium oxide and tellurium tetrachloride ($LiCl-Li_2O-TeCl_4$) at 650 °C for a duration ranging from 30 to 120 min under a galvanostatic mode. The tellurium coating was observed to be thick, uniform, smooth and homogeneous. Additionally, the deposited tellurium did not chemically react with the iridium substrate to form intermetallic compounds, which is a good feature from the standpoint of the device's performance characteristics. The present process, being generic in nature, shows the potential for the manufacture of both the coated substates and high-purity elements not just for tellurium but also for other p-block elements.

Keywords: molten salt; three-electrode set up; galvanostatic deposition; p-block elements; coated component

1. Introduction

P-block elements, ranging from group 13 through 18 except helium from group 18 in the periodic table, consist of metals, nonmetals and semimetals/metalloids. Some of these elements, such as silicon, selenium, tellurium, bismuth, indium, gallium, germanium and their compounds/alloys, are extensively used in many key technologies including the clean energy sector. Due to an increased demand for photovoltaic cells, some of these elements have become the most sought-after commodities for the manufacture of these devices/modules to capture solar energy in a cost-effective manner. At present, silicon (both in crystalline and amorphous forms) dominates the photovoltaic market. Other forms of solar modules (thin film/perovskite/organic/quantum dots/multijunction/concentration photovoltaics) are gradually emerging in the scene because of their inexpensiveness, versality and relatively better efficiencies. In order to cater to the increasing demand, the recycling of waste photovoltaic modules to recover precious metals, such as germanium, selenium, tellurium and remove toxic elements (cadmium, lead) has, in recent years, gained global momentum [1–3].

From a device-fabrication standpoint, these elements are mostly preferred as coatings on a suitable substrate, such as metals, glass or plastics, and are routinely used as semiconductors, photovoltaics, phase change memory materials, high performance thermoelectric materials and alloy-additive components/films [4,5]. Tellurium, being one of the p-block elements, is widely used in photovoltaic modules [6]. Coating of tellurium has

Citation: Tripathy, P.K.; Mondal, K. A Molten Salt Electrochemical Process for the Preparation of Cost-Effective p-Block (Coating) Materials. *Crystals* **2022**, *12*, 385. https://doi.org/10.3390/cryst12030385

Academic Editors: Radu Claudiu Fierascu and Florentina Monica Raduly

Received: 24 February 2022
Accepted: 11 March 2022
Published: 13 March 2022

Publisher's Note: MDPI stays neutral with regard to jurisdictional claims in published maps and institutional affiliations.

Copyright: © 2022 by the authors. Licensee MDPI, Basel, Switzerland. This article is an open access article distributed under the terms and conditions of the Creative Commons Attribution (CC BY) license (https://creativecommons.org/licenses/by/4.0/).

been traditionally carried out by multiple techniques, such as physical vapor deposition (PVD), chemical vapor deposition (CVD), ion implantation, atomic layer deposition (ALD) and electrodeposition [7–10]. Unlike the vacuum-assisted manufacturing methods (which are essentially top-down processes with several flow control/adjustment attachments), electrodeposition processes offer a one-pot and bottom-up synthesis approach whereby all the precursor components are present in one single solution. As a result of the inherent advantages associated with the electrodeposition techniques, these processes have been widely used in the fabrication of advanced engineering materials including metals, alloys, compounds, semimetals, non-metals and thin films.

A number of studies on the electrodeposition of p-block elements, their alloys and compounds including tellurium, from aqueous-based baths have shown limited success. Some of these drawbacks include (i) the low solubility of the functional electrolytes (compounds containing silicon, tellurium, selenium) in water/organic electrolytes (ii) non-adherence of the films/coatings on the substrate materials (iii) solubility of the electrodeposited material/coating in the electrolyte and (iv) formation of relatively thin deposits/coatings. For example, tellurium coatings prepared from aqueous solutions including ionic liquids, were reported to be of inferior quality [11–13]. These coatings, when prepared from a ternary chloride melt ($AlCl_3$-NaCl-KCl, using $TeCl_4$ as the functional electrolyte), have also been observed to suffer from similar inadequacies. The electrochemical behavior of tellurium in the ternary melt has been reported to be complex in nature [12]. Moreover, the electrodeposited layers were observed to be non-adherent and dissolved into the melt immediately after formation [12]. Another example is aluminum-coated steels, prepared from low temperature chloride melts in the temperature range of 160–180 °C. These coatings were also observed to be of non-adherent type besides not being able to show good corrosion-resistance abilities. [13,14]. It appears that a relatively elevated temperature may remove some of the limitations that are inherent to low-temperature coating/deposition processes. A recent study from our laboratory, has indeed shown the superior features of the coated components in terms of good adhesion and excellent corrosion-resistance properties in a simulated marine environment [15]. Several published literatures also support such a viewpoint wherein elevated temperatures have been shown to promote formation of smooth, adherent, thick and relatively uniform deposits/coatings [16,17]. One of the key factors, among others, that contribute to the formation of relatively thicker electrodeposits/coatings can be ascribed to the increased solubility of the functional electrolyte. Unlike aqueous electrolytes and ionic liquids, molten salts are endowed with this property as these can operate at elevated temperatures.

The present study describes results from one of our recent experimental research efforts whereby the focus was to examine the suitability of a platinum group metal (PGM) for its deployment as an oxygen-evolving electrode during the electrochemical reduction of used uranium oxide. The specific objective of the research was to study the chemical interactions of a group of reactive gaseous elements (oxygen, selenium, tellurium and iodine) that were made to generate electrochemically, on the performance characteristics as well as mechanical integrity of the PGM. Platinum, as an inert anode, is known to undergo degradation during the electrochemical reduction of uranium oxide [18]. The specific objective of the present experiment was to replace platinum with another PGM, namely iridium, to examine its interaction with tellurium. In situ generated tellurium, via the electrochemical reduction of tellurium tetrachloride ($TeCl_4$), was deposited on the iridium to examine its chemical interaction with iridium.

2. Experimental

The electrochemical runs were carried out in an argon atmosphere (containing < 0.1 ppm of moisture and oxygen) by placing the electrochemical cell (Figure 1) in the glove box. High purity and a combination of ultra-dry LiCl-1.0 wt.% Li_2O was used as the supporting electrolyte. Anhydrous $TeCl_4$ (equivalent to 0.25 wt.% tellurium) was used as the source of tellurium (functional electrolytes). A three-electrode setup, consisting of an iridium wire

(1 mm dia., 99.8% pure), molybdenum coil (made from 1 mm dia., 99.9% pure Mo wire) and glassy carbon (3 mm dia. and 100 mm long) as working, counter and reference electrodes, respectively (Figure 2), was used to carry out the electrodeposition test runs. The electrodes were sheathed with high purity alumina tubes to insulate them (from establishing possible electrical contacts with the electrochemical cell). A magnesia crucible was used to contain the molten electrolyte.

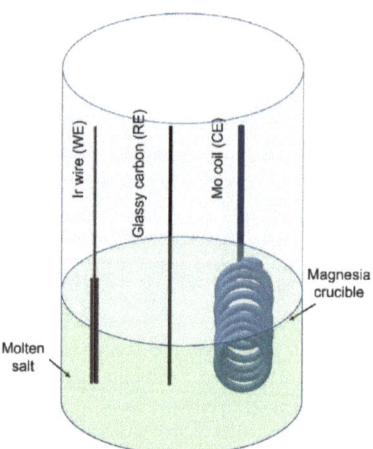

Figure 1. Schematic diagram of the electrochemical cell assembly.

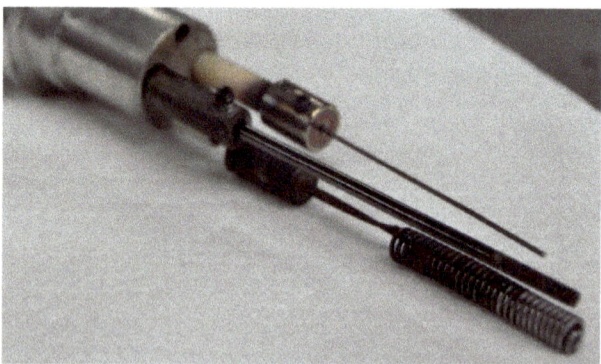

Figure 2. Three-electrode set up (iridium wire as the working electrode; glassy carbon as the pseudo reference electrode and molybdenum coil as the counter electrode).

The electrochemical cell was assembled, electrodes were placed above the crucible (in a vertical position) and the furnace was switched on to the heating mode in a controlled manner. The temperature was set at 650 °C and the measured temperature, during electrodeposition tests, was observed to be 650 ± 2 °C. The melt was allowed to homogenize for about an hour prior to the electrochemical measurements. The electrodes were lowered into the melt and electrical connections were made via the Biologic potentiostat-galvanostat to record the experimental data. The electrodeposition runs were carried out, at constant current mode (0.5–2.0 A) and at 650 °C, using a suite of transient electrochemical techniques. The coated samples were evaluated and characterized by XRD and ICP-MS measurements.

3. Results and Discussion

3.1. Cyclic Voltammetric Measurements

A series of cyclic voltammetric (CV) measurements, at different scan rates (0.025–0.15 V s^{-1}), were performed to determine the decomposition voltage of the TeCl$_4$. The CV data showed consistency (in the measured values of the decomposition voltage), both during the cathodic deposition of tellurium (on iridium) and anodic stripping of tellurium (back to the electrolyte). Figure 3 shows a typical CV indicating the deposition of tellurium (during cathodic sweep) and its dissolution (in the electrolyte) during the reverse (also known as anodic) sweep. The appearance of a single redox peak indicated that the deposition of tellurium and its subsequent stripping occurred in one step. The nature of the CV plot indicated the electrode kinetics to be reversible and diffusion controlled.

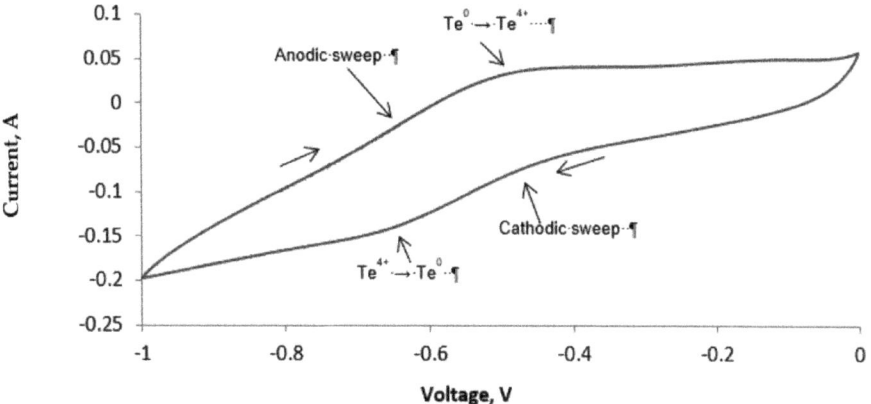

Figure 3. Cyclic voltammogram (CV) of TeCl$_4$ in LiCl-Li$_2$O-TeCl$_4$ electrolyte, [Te^{4+}] –0.25 wt.%, S = 0.5 cm^2, T –650 °C.

The appearance of a single peak, during the square wave (SQV) measurement, followed by the CV measurements, further confirmed the fact that the tellurium deposition and dissolution occurred in just one step (Figure 4).

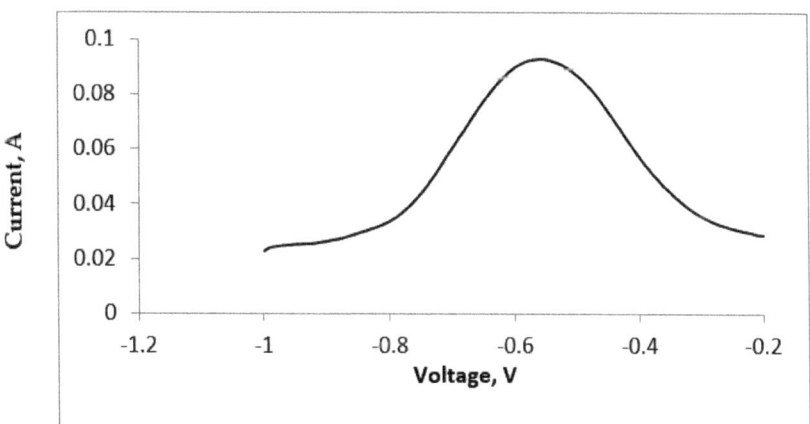

Figure 4. Square Wave Voltammogram of the TeCl$_4$: Pulse height −25 mV, potential step −10 mV, Frequency −8 Hz.

3.2. Galvanostatic Experiment

After experimentally determining the decomposition voltage of TeCl$_4$, both by CV and SQV electrolysis experiments, were performed at constant current to deposit tellurium onto the iridium cathode. The galvanostatic experiments were carried out both at 0.1 A and then at 0.2 A for a total duration ranging between 30–120 min. A steady voltage (after an initial instability) indicated the deposition of tellurium into iridium (Figure 5).

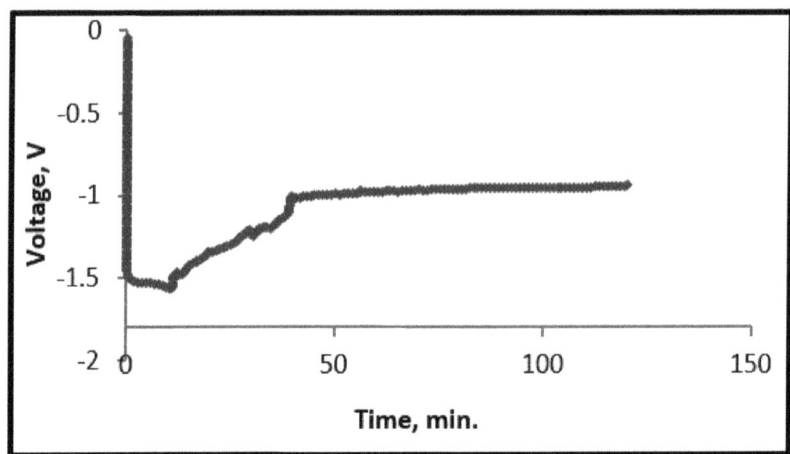

Figure 5. Voltage vs. time profile: current −0.2 A and duration −120 min.

The deposition run was stopped after 120 min to allow the furnace to cool to room temperature. The iridium wire was subsequently taken out of the electrochemical cell for further evaluation and characterization. The iridium wire, after the removal from the electrochemical cell, was observed to have been coated with a smooth, adherent, uniform and thick tellurium (surface) coating (Figure 6).

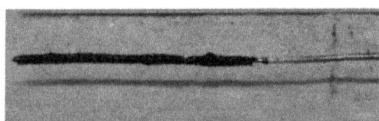

Figure 6. Tellurium-coated iridium wire (before washing/cleaning).

3.3. Inductively Coupled Plasma–Mass Spectrometric (ICP-MS) Analysis of the Deposit

The deposit (Figure 6) was scraped with a knife to analyze its chemical composition (without washing). As expected, the analysis indicated the presence of two major elements (tellurium and lithium) (Table 1).

Table 1. ICP-MS analysis of the scraped coating.

Element	Amount (wt.%)
Li	75.8
Te	24.2
Ir	0.1

3.4. XRD Patterns of the Deposits

A small portion of the unwashed deposit was cut and subjected to the XRD to determine the phase compositions. The XRD pattern indicated the formation of four distinct phases, as follows: elemental tellurium, two lithium telluride phases, $LiTe_3$, Li_2Te, respectively and elemental iridium. The XRD did not reveal the formation of any iridium telluride phase(s) (Ir_xTe_y). The coated wire was subsequently washed with distilled water to remove the salt and analyze the integrity of surface tellurium on iridium. The XRD (Figure 7) indicated the presence of tellurium as the major phase with traces of lithium and iridium as the minor phase.

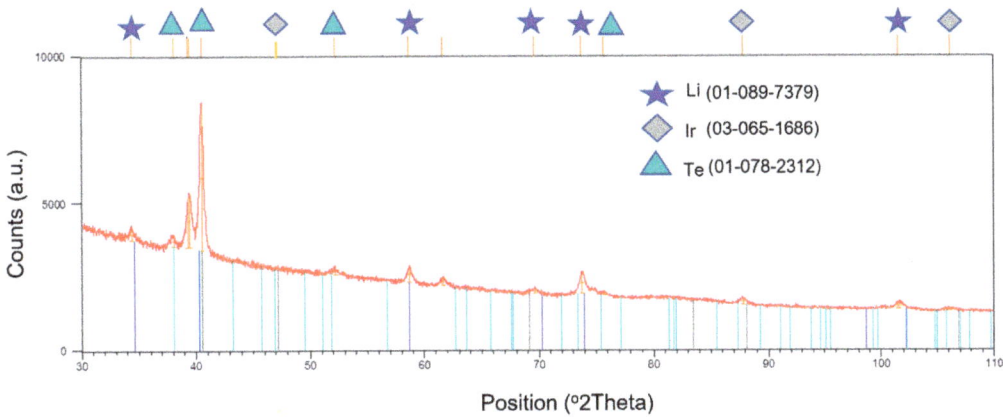

Figure 7. XRD pattern of the washed tellurium-coated iridium substrate/wire.

Although thermodynamics favors the formation of two iridium tellurides ($IrTe_2$ and $IrTe_{2.67}$, respectively) [19], the XRD pattern did not indicate their formation. Perhaps their formation was inhibited due to kinetic considerations. These results have led to the following conclusions:

1. Unlike platinum, the mechanical integrity of iridium was not observed to have been adversely impacted. Iridium and tellurium did not form any brittle telluride phases. These studies have clearly indicated the superior features of iridium over platinum.
2. The tellurium deposit was observed to be highly pure.
3. The tellurium coating on iridium was observed to be smooth, thick and adherent. The coating did not undergo any spallation and/or form cracks even after repeated washing.
4. Iridium can be used as a metal substrate to form tellurium-coated components for solar energy devices.

4. Conclusions

Present experimental studies have conclusively established the feasibility of preparing p-block elements, such as tellurium, via an inexpensive molten-salt electrochemical process. Besides being cost-effective in nature, the electrochemical process shows five tangible benefits: (i) preparation of pure elements (ii) possibility of the fabrication of in situ coated components for direct use in a device (iii) easy control of deposition parameters to control the tellurium thickness on the metallic substrates (iv) generic nature of the molten salt electrochemical process for the fabrication of other p-block elements/coatings and (v) superior features of iridium as a potential metallic substrate material.

Author Contributions: Conceptualization, P.K.T.; methodology, K.M. All authors have read and agreed to the published version of the manuscript.

Funding: No external funding was received.

Institutional Review Board Statement: The institutional approval number is: INL/JOU-21-64989.

Informed Consent Statement: Not applicable.

Data Availability Statement: Not applicable.

Acknowledgments: The work was supported by the Idaho National Laboratory Directed Research and Development Program under DOE Idaho Operations Office. The manuscript was authorized by Battelle Energy Alliances under the contract No. DE-AC07-05ID14517, with the US Department of Energy, for publication. The US government retains and the publisher, by accepting the manuscript for publication, acknowledges that the US government retains a non-exclusive, paid up irrevocable worldwide license to publish or reproduce the published form of this manuscript or allow others to do so for United States Government purposes.

Conflicts of Interest: The authors declare that they have no known competing financial interests or personal relationships that could have appeared to influence the work reported in this paper.

References

1. Smith, Y.R.; Nagel, J.R.; Rajamani, R.K. Electrodynamic Eddy Current Separation of End-of-Life PV Materials. In *Energy Technology 2017*; The Minerals, Metals & Materials Series; Zhang, L., Ed.; Springer: Berlin/Heidelberg, Germany, 2017; pp. 379–386.
2. Bogust, P.; Smith, Y.R. Physical separation and beneficiation of end-of-life photovoltaic panel materials: Utilizing tem-perature swings and particle shape. *JOM* **2020**, *72*, 2615–2623. [CrossRef]
3. Bruckman, L.S. Transformative Opportunities from Data Science and Big Data Analytics: Applied to Photovoltaics. *Electrochem. Soc. Interface* **2019**, *28*, 57–61. [CrossRef]
4. Bartlett, P.N.; Cook, D.; de Groot, C.H.; Hector, A.L.; Huang, R.; Jolleys, A.; Kissling, G.P.; Levason, W.; Pearce, S.J.; Reid, G. Non-aqueous electrodeposition of p-block metals and metalloids from halometallate salts. *RSC Adv.* **2013**, *3*, 15645–15654. [CrossRef]
5. Kowalik, R.; Kutyła, D.; Mech, K.; Tokarski, T.; Zabinski, P. Electrowinning Of Tellurium From Acidic Solutions. *Arch. Met. Mater.* **2015**, *60*, 591–596. [CrossRef]
6. Redlinger, M.; Eggert, R.; Woodhouse, M. Evaluating the availability of gallium, indium, and tellurium from recy-cled photovoltaic modules. *Sol. Energy Mater. Sol. Cells* **2015**, *138*, 58–71. [CrossRef]
7. Sen, S.; Bhatta, U.M.; Kumar, V.; Muthe, K.P.; Bhattacharya, S.; Gupta, S.K.; Yakhmi, J.V. Synthesis of Tellurium Nanostructures by Physical Vapor Deposition and Their Growth Mechanism. *Cryst. Growth Des.* **2008**, *8*, 238–242. [CrossRef]
8. Ma, Y.-T.; Gong, Z.-Q.; Xu, W.-H.; Huang, J. Structural and optical properties of tellurium films obtained by chemical vapor deposition(CVD). *Tran. Nonferr. Met. Soc. China* **2006**, *16*, 693–699. [CrossRef]
9. Kowalik, R.; Kutyła, D.; Mech, K.; Żabiński, P. Analysis of tellurium thin films electrodeposition from acidic citric bath. *Appl. Surf. Sci.* **2016**, *388*, 817–824. [CrossRef]
10. Johnson, R.W.; Hultqvist, A.; Bent, S.F. A brief review of atomic layer deposition: From fundamentals to applications. *Mater. Today* **2014**, *17*, 236–246. [CrossRef]
11. Ito, S.; Kitagawa, N.; Shibahara, T.; Nishino, H. Electrochemical Deposition of Te and Se on Flat TiO_2 for Solar Cell Application. *Int. J. Photoenergy* **2014**, *2014*, 5. [CrossRef]
12. Ebe, H.; Ueda, M.; Ohtsuka, T. Electrodeposition of Sb, Bi, Te, and their Alloys in $AlCl_3$–NaCl–KCl Molten Salt. *Electrochim. Acta* **2007**, *53*, 100–105. [CrossRef]
13. Fellner, P.; Chrenkova-Paucirova, M.; Matiasovsky, K. Electrolytic Aluminum Plating in Molten Salt Mixtures Based on $AlCl_3$ I: Influence of the Addition of Tetra methyl ammonium chloride. *Surf. Technol.* **1981**, *14*, 101–108. [CrossRef]
14. Paucirova, M.; Matiasovsky, K. Electrolytic Aluminum-Plating in Fused Salts Based on Chlorides. *Electrodepos. Surf. Treat.* **1975**, *3*, 121–128. [CrossRef]
15. Tripathy, P.K.; Wurth, L.A.; Dufek, E.J.; Gutknecht, T.Y.; Gese, N.J.; Hahn, P.A.; Frank, S.M.; Fredrickson, G.L.; Herring, J.S. Aluminum electroplating on steel from a fused bromide electrolyte. *Surf. Coat. Technol.* **2014**, *258*, 652–663. [CrossRef]
16. Wu, G.; Li, N.; Zhou, D.; Mitsuo, K. Electrodeposited Co–Ni–Al_2O_3 composite coatings. *Surf. Coat. Technol.* **2004**, *176*, 157–164. [CrossRef]
17. Gu, Y.; Liu, J.; Qu, S.; Deng, Y.; Han, X.; Hu, W.; Zhong, C. Electrodeposition of alloys and compounds from high-temperature molten salts. *J. Alloys Compd.* **2017**, *690*, 228–238. [CrossRef]
18. Herrmann, S.D.; Tripathy, P.K.; Frank, S.M.; King, J.A. Comparative Study of Monolithic Platinum and Iridium as Oxy-gen-evolving Anodes during the Electrolytic Reduction of Uranium Oxide in a Molten $LiCl$-Li_2O Electrolyte. *J. Appl. Electrochem.* **2019**, *49*, 379–388. [CrossRef]
19. *H.S.C. Chemistry, Ver. 7*; Outotec: Pori, Finland, 2015.

Article

Modification of $FA_{0.85}MA_{0.15}Pb(I_{0.85}Br_{0.15})_3$ Films by NH_2-POSS

Yangyang Zhang [1,†], Na Liu [2,†], Haipeng Xie [1,*], Jia Liu [1], Pan Yuan [1], Junhua Wei [1], Yuan Zhao [1], Baopeng Yang [1], Jianhua Zhang [1], Shitan Wang [1], Han Huang [1], Dongmei Niu [1], Qi Chen [2] and Yongli Gao [3]

1. Institute of Super-Microstructure and Ultrafast Process in Advance Materials, School of Physics and Electronics, Central South University, Changsha 410012, China; 192212082@csu.edu.cn (Y.Z.); jialiu1@csu.edu.cn (J.L.); yp123217@csu.edu.cn (P.Y.); wjh101@csu.edu.cn (J.W.); zhaoyuan@csu.edu.cn (Y.Z.); baopengyang@csu.edu.cn (B.Y.); zhangjianhua@csu.edu.cn (J.Z.); shitan2021@163.com (S.W.); physhh@csu.edu.cn (H.H.); mayee@csu.edu.cn (D.N.)
2. Beijing Key Laboratory of Nanophotonics and Ultrafine Optoelectronic Systems, School of Materials Science & Engineering, Beijing Institute of Technology, Beijing 100081, China; 3120170530@bit.edu.cn (N.L.); chacha.chenqi@gmail.com (Q.C.)
3. Department of Physics and Astronomy, University of Rochester, Rochester, NY 14627, USA; ygao@pas.rochester.edu
* Correspondence: xiehaipeng@csu.edu.cn
† These authors contributed equally to this work.

Abstract: The surface composition and morphology of $FA_{0.85}MA_{0.15}Pb(I_{0.85}Br_{0.15})_3$ films fabricated by the spin-coating method with different concentrations of NH_2-POSS were investigated with atomic force microscopy (AFM), X-ray photoelectron spectroscopy (XPS), angle-resolved X-ray photoelectron spectroscopy (AR-XPS), and Fourier transform infrared spectroscopy (FTIR). It was found that the surface composition of the $FA_{0.85}MA_{0.15}Pb(I_{0.85}Br_{0.15})_3$ films was changed regularly through the interaction between NH_2-POSS and the perovskite film. The corresponding surface morphological changes were also observed. When the concentration of NH_2-POSS exceeded 10 mg/mL, a lot of cracks on the surface of the perovskite film were observed and the surface morphology was damaged. The surface composition and its distribution can be adjusted by changing the concentration of NH_2-POSS and the proper concentration of NH_2-POSS can substantially improve the quality of perovskite film.

Keywords: perovskite film; NH_2-POSS; surface composition; morphology; photoemission spectroscopy

1. Introduction

Hybrid organic-inorganic perovskite (HOIP) semiconductors have attracted a lot of intensive interest recently owing to their excellent advantages, such as a large absorption coefficient, long charge carrier diffusion lengths, and distinguished optoelectronic properties [1–8]. Recently, a power conversion efficiency (PCE) of 25.5% was certified by the National Renewable Energy Laboratory [9]. It is known that the stability of perovskite films is poor in wet environments, which leads to a decrease of the efficiency decrease or even failure of perovskite solar cells (PSCs). In order to obtain high-performance PSCs, a lot of studies have focused on interface engineering, such as optimizing the transmission layer material, interface control, and surface modification [10–15]. Among them, surface modification (e.g., metal ion doping and modification of common chemical additives) is one of the effective methods to improve the performance of organic-inorganic hybrid perovskite devices.

Zhou et al. explored an effective approach for simultaneous passivation of cation and anion vacancy defects in perovskite materials by adding NaF and achieved a power conversion efficiency of 21.46% [16]. Liu et al. found that a power conversion efficiency of 20.13% was realized by doping different proportions of Cs^+ in the $(CH_3NH_3)_{1-x}Cs_xPbI_3$ perovskite film [17]. Seok et al. demonstrated that the introduction of additional iodide ions into the organic cation solution can reduce the concentration of deep defects and achieved a power conversion efficiency of 22.1% [18]. Marco et al. reported that the carrier lifetimes and open circuit voltages in hybrid perovskite solar cells can be enhanced via introducing guanidinium-based additives [19]. Snaith et al. reported that treating the crystal surfaces with the Lewis bases thiophene and pyridine can reduce nonradiative electron-hole recombination in the $CH_3NH_3PbI_{3-x}Cl_x$ perovskite film [20]. Park et al. found that adding a tiny amount of potassium iodide to perovskite materials can greatly reduce current-voltage hysteresis and the potassium ion was best for preventing Frenkel defects as compared to other alkali metal cations [21].

Here, we introduce a new chemical additive, namely, polyhedral oligomeric silsesquioxane (POSS), which is non-toxic and tasteless, and has good biocompatibility, stability, and other characteristics. The POSS molecules have a rigid cage-like stereo structures with nanometer sizes and it can be used to prepare new organic-inorganic hybrid materials [22–25]. For example, the aminopropyllsobutyl POSS (NH_2-POSS) has been used to passivate the surface of perovskite films due to its excellent chemical stability and optical properties [26]. Because the $FA_{0.85}MA_{0.15}Pb(I_{0.85}Br_{0.15})_3$ polycrystalline film has obvious defects at the grain boundaries, Chen et al. reported that the NH_2-POSS passivation at the surface and grain boundaries of perovskite film can significantly reduce the trap density and trap state energy level, which in turn enhances the PCE from 18.1% to 20.5% [27]. Some fundamental problems are still unclear, such as the change of the surface composition and the distribution of chemical components in perovskite films modified by different NH_2-POSS concentrations.

In this study, surface modification of the $FA_{0.85}MA_{0.15}Pb(I_{0.85}Br_{0.15})_3$ films was carried out by introducing the chemical additive NH_2-POSS. X-ray, angle-resolved X-ray photoelectron spectroscopy (XPS, AR-XPS), atomic force microscopy (AFM), and Fourier transformed infrared spectroscopy (FTIR) were used to study the effects of NH_2-POSS on the surface composition, component distribution, and surface morphology of perovskite films. It was found that NH_2-POSS not only changes the surface composition and its distribution, but also modifies the surface morphology of the perovskite films due to the interaction between NH_2-POSS and the perovskite. The quality of the perovskite film can be improved by using a proper concentration of NH_2-POSS. The surface uniformity of the perovskite film is damaged when the concentration of NH_2-POSS is above 10 mg/mL.

2. Experimental

The $FA_{0.85}MA_{0.15}Pb(I_{0.85}Br_{0.15})_3$ films and NH_2-POSS were prepared according to the reported procedure [27]. The PbI_2 solution was deposited by spin coating at 2000 rpm for 30 s and dried at 70 °C for 30 min. After the PbI_2-coated substrate was cooled to room temperature (25 °C) in a nitrogen glovebox, a solution composed of $HC(NH_2)_2I$, CH_3NH_3I, CH_3NH_3Br, and CH_3NH_3Cl was deposited by spin coating at 3000 rpm for 30 s in a nitrogen glovebox and dried at 150 °C for 15 min in air. The NH_2-POSS solution in chlorobenzene was dropped onto the perovskite substrate, immediately followed by spin coating at 3000 rpm for 30 s, and then annealing at 100 °C for 5 min to completely remove the solvent.

The $FA_{0.85}MA_{0.15}Pb(I_{0.85}Br_{0.15})_3$ films modified by different concentrations of NH_2-POSS were sent into the characterization chamber for photoelectron spectroscopy measurements. The XPS was measured with a monochromatic Microfocus X-ray Source (Al K_α, hv = 1486.7 eV). For XPS, a resolution of 0.65 eV of the spectrometer was chosen with a pass energy of 40 eV. The binding energies of all XPS were calibrated and referenced to the Femi level (E_F) of the sample [28–31]. The morphology of the perovskite films was measured ex situ using an Agilent 5500AFM/SPM system. The structures of perovskites with/without NH_2-POSS were characterized by a Fourier transform infrared spectrophotometer (Nicolet™ iS™ 50 FTIR Spectrometer; Thermo Fisher, Waltham, MA, USA) using the KBr wafer technique. All measurements were taken at room temperature.

3. Results and Discussion

The chemical structure of NH_2-POSS is shown in Figure 1a, where the Si, O, H, C, and N atoms are represented by brownish-red, cyan, white, gray, and red balls, respectively. The NH_2-POSS was introduced into perovskite films through a spin-coating method and the XPS full scan spectra of the perovskite films fabricated with different concentrations of NH_2-POSS are shown in Figure 1b,c. The Si 2p peak appeared at about 103.81 eV when the concentration of NH_2-POSS was 3 mg/mL, and the intensity of Si 2p changed consistently as the concentration of NH_2-POSS was increased. In addition, the FTIR spectrum shows that the peak located at about 1103 cm^{-1} corresponds to the asymmetric stretching vibration of Si-O-Si [32] (see the Supplementary Materials Figure S1). This indicates that the NH_2-POSS was successfully introduced into the surface of the perovskite films. It is also remarkable that a slight presence of the oxygen component was detected in the XPS full-scan spectra of the perovskite film without NH_2-POSS modification, which can be attributed to contamination from the ex situ preparation of the sample [33].

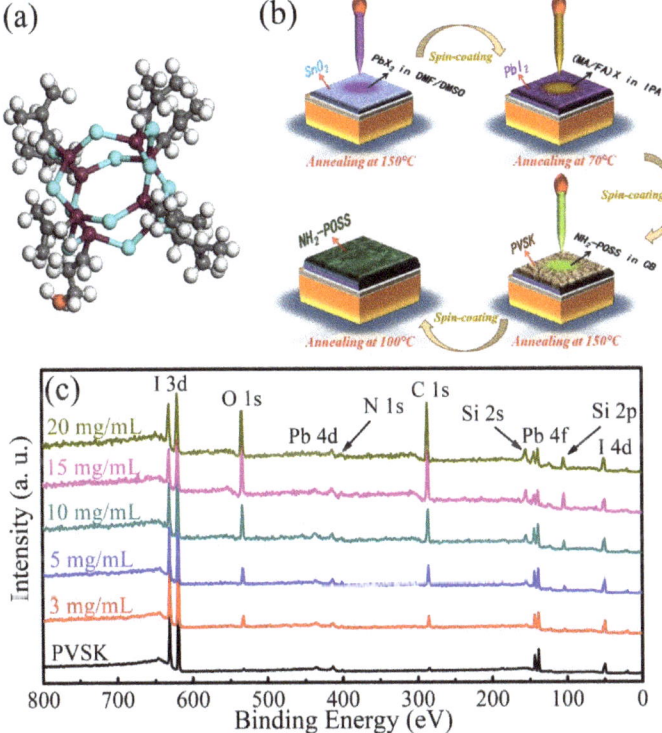

Figure 1. (a) The chemical structure of NH_2-POSS. The presence of Si features is from NH_2-POSS. (b) The schematic scheme of the surface modification process with NH_2-POSS. (c) The XPS full-scan spectra of the perovskite films fabricated with different concentrations of NH_2-POSS.

We further analyzed the element compositions of perovskite films before and after modification by NH_2-POSS. In Figure 2, the I $3d_{5/2}$, Pb $4f_{7/2}$, and N 1s core level of the perovskite films fabricated with different concentrations of NH_2-POSS are presented. For visual clarity, all the spectra were normalized to the same height. As shown in Figure 2a, the I $3d_{5/2}$ core level is located at about 619.55 eV (I1) for the as-grown perovskite film, corresponding to the I component in the perovskite film. After modification by NH_2-POSS, we observed that a new peak at about 620.74 eV (I2) appeared at the high binding energy. As the concentration of NH_2-POSS increased from 3 to 20 mg/mL, the position of the I1 and I2 peak was unchanged. Similar to the I element, the Pb $4f_{7/2}$-related peaks located at about 138.78 eV (Pb1) can be attributed to the Pb component in the perovskite film, and a new peak located at about 140.23 eV (Pb2) was observed after modification by NH_2-POSS, as shown in Figure 2b. Comparing Figure 2a with Figure 2b, it is found that the intensity of the I2 and Pb2 peaks increased with the increase of the NH_2-POSS concentration. This indicates that NH_2-POSS is attached to the perovskite film surface via the interaction between the amino anchoring group and the Pb-I lattice [20,34]. In Figure 2c, the N 1s core level includes more than one peak before and after modification by NH_2-POSS for all perovskite films. For the as-grown perovskite film, the N 1s core level located at about 400.94 (N1) and 402.77 eV (N2) can be assigned to the N element in FA and MA, respectively. It is found that the intensity of N2 increases after modification by NH_2-POSS, which can be attributed to the introduction of the N element in NH_2-POSS. In addition, the full width at half maxima (FWHM) of N1 remained basically unchanged as the NH_2-POSS concentration was increased from 0 to 20 mg/mL, while the FWHM of N2 increased from 1.14 to 2.84 eV (see the Supplementary Materials Table S1). This could be due to the interaction between NH_2-POSS molecules and perovskite films.

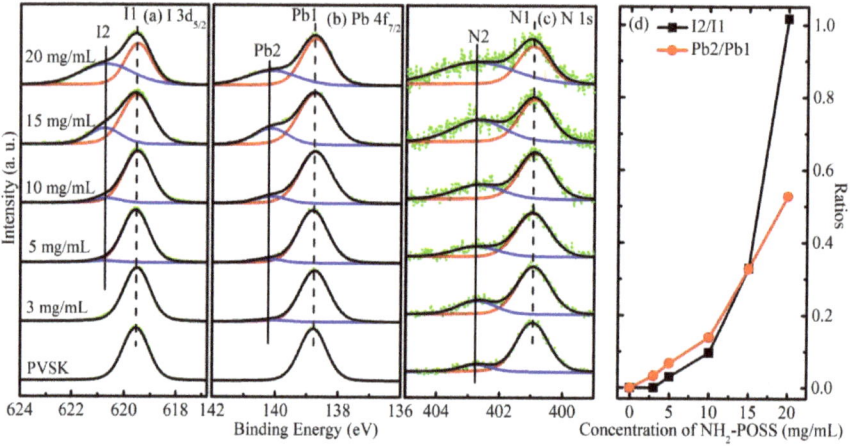

Figure 2. XPS spectra of (**a**) I $3d_{5/2}$, (**b**) Pb $4f_{7/2}$, and (**c**) N 1s for perovskite films fabricated with different concentrations of NH_2-POSS. (**d**) The ratio of I2/I1 and Pb2/Pb1 with different concentrations of NH_2-POSS.

As shown in Figure 2d, the ratio of I2/I1 and Pb2/Pb1 increases as the concentration of NH_2-POSS increased and it can be divided into two different stages. In the first stage, the ratio of I2/I1 and Pb2/Pb1 increases linearly and slowly before the concentration of NH_2-POSS reaches 10 mg/mL. The same trend of the I2/I1 ratio and Pb2/Pb1 ratio indicated an interaction between NH_2-POSS and perovskite film during the first stage, which agrees with previously reported results [27,34]. In the second stage, the ratio of I2/I1 and Pb2/Pb1 increases quickly with further increasing of the concentration of NH_2-POSS, indicating that the interaction between NH_2-POSS and perovskite film was changed. It is further proven that the compositions of the perovskite films can be affected by NH_2-POSS.

As shown in Figure 3, the atomic concentrations of all components were plotted as a function of the concentration of NH_2-POSS. It is observed that the atomic concentration of I1, Pb1, N1, and C3 decreases as the concentration of NH_2-POSS increases, which can be attributed to the surface modification with NH_2-POSS. As shown in Figure 3a,b, the atomic concentration of I1 is about 12% and 6% at 0° and 60° (take-off angle), respectively, when the concentration of NH_2-POSS is 3 mg/mL. It is the same for the components Pb1, N1, and C3, which come from the perovskite film. In Figure 3b, the atomic concentration of Si is about 23% at 3 mg/mL NH_2-POSS. It increases to about 25% at 5 mg/mL and then slightly increases as the concentration of NH_2-POSS increases, which is the same for C1, Si, and O from NH_2-POSS. Otherwise, the content of I2 and Pb2 increases slowly as the concentration of NH_2-POSS increases, as shown in Figure 3a,b. It indicates that there is interaction between NH$_2$-POSS and the perovskite film, mostly on the surface of the perovskite film.

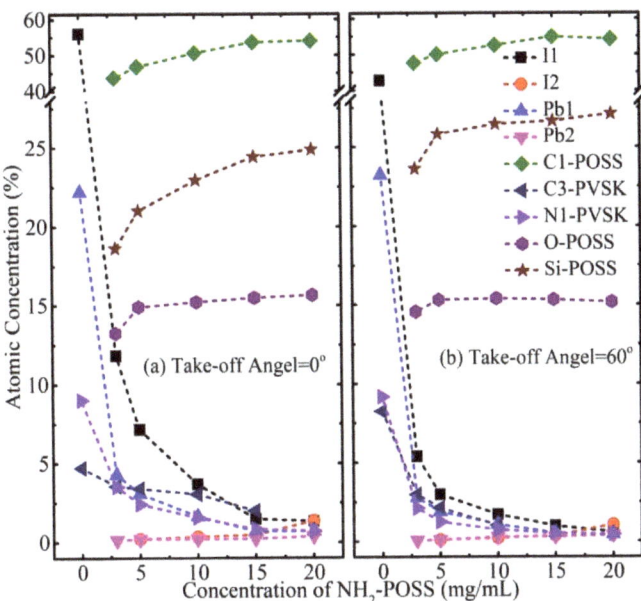

Figure 3. Atomic concentration of all components in the perovskite films as a function of the concentration of NH$_2$-POSS. (**a**) The take-off angle is 0°; (**b**) the take-off angle is 60°.

The relative depth plot of the chemical components of perovskite films fabricated with different concentrations of NH$_2$-POSS is shown in Figure 4. For the as-grown perovskite film, it is observed that the C3 component has a smaller relative depth, the I1 component has a grater relative depth, and the N1 and Pb1 components have a moderate relative depth, as shown in Figure 4a. After modification by NH$_2$-POSS, the Si, O, and C1 components have a smaller relative depth; and the Pb2 component has a greater relative depth when the concentration of NH$_2$-POSS is 3 or 5 mg/mL, as shown in Figure 4b,c. With a further increase of the concentration of NH$_2$-POSS, the Pb2, Si, C1, and O components have a smaller relative depth. When the concentration of NH$_2$-POSS reaches 10 mg/mL, the relative depth of Pb2 changes significantly, as shown in Figure 4d. This indicates that the interfacial reaction between NH$_2$-POSS and superficial perovskite is complete and the Pb2 is mostly on the surface. As the take-off angle reaches 60°, the content of Pb2 remains basically unchanged and that of Pb1 decreases, so Pb2 has a smaller relative depth. The relative depth of Pb1, I2, N1, and I1 remains almost unchanged at the concentration of 10 mg/mL. This can be attributed to the interaction between the perovskite film and NH$_2$-POSS, which has an influence on the surface composition and its structure.

Figure 5 shows the surface morphologies of the perovskite films fabricated with different concentrations of NH$_2$-POSS investigated using AFM. The height profiles correspond to the height variation of the blue line in the AFM images. It can be clearly observed that the surface morphologies of the perovskite films are changed by NH$_2$-POSS. The morphology of the as-grown perovskite film is shown in Figure 5a. Crystalline particles are found on the surface, and the corresponding root mean square (RMS) is 24.87 nm. After the modification by NH$_2$-POSS, the surface morphologies of the perovskite films were changed significantly and no crystalline particle was observed on the surface, as shown in Figure 5b–f. When the concentration of NH$_2$-POSS is 10 mg/mL, the RMS is about 12.43 nm, which is the lowest value of all modified perovskite films. It is noticeable that there are a lot of cracks on the surface of the perovskite films, and the height profile is obviously concave downward when the concentration of NH$_2$-POSS exceeds 10 mg/mL, as shown in Figure 5e,f. It suggests that the NH_2-POSS can damage the surface morphology when the concentration of NH_2-POSS is 15 and 20 mg/mL, which will lead to a decrease of the device's performance. As a matter of fact, the best device performance was obtained after modification with 10 mg/mL NH_2-POSS in a previous study [27].

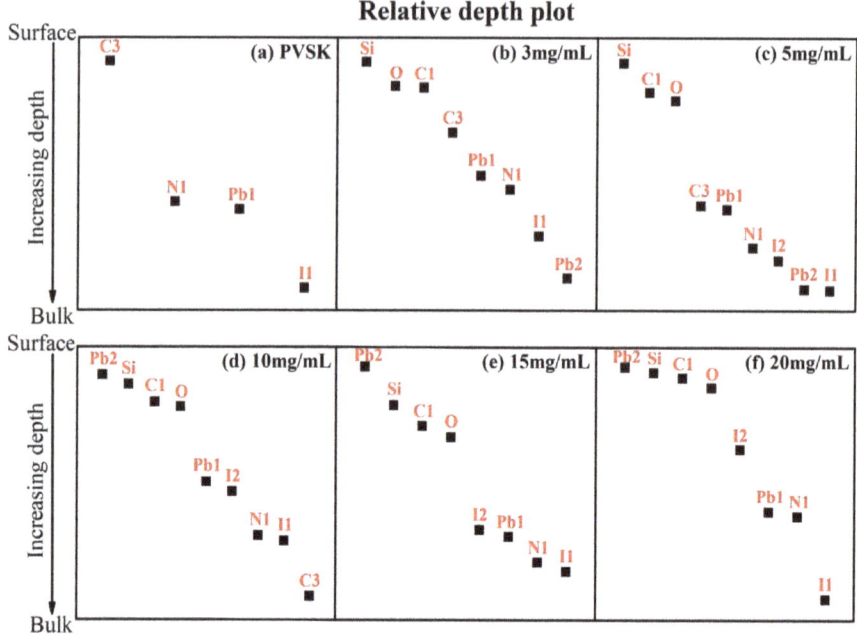

Figure 4. Relative depth plot of the chemical components of perovskite films fabricated with different concentrations of NH_2-POSS.

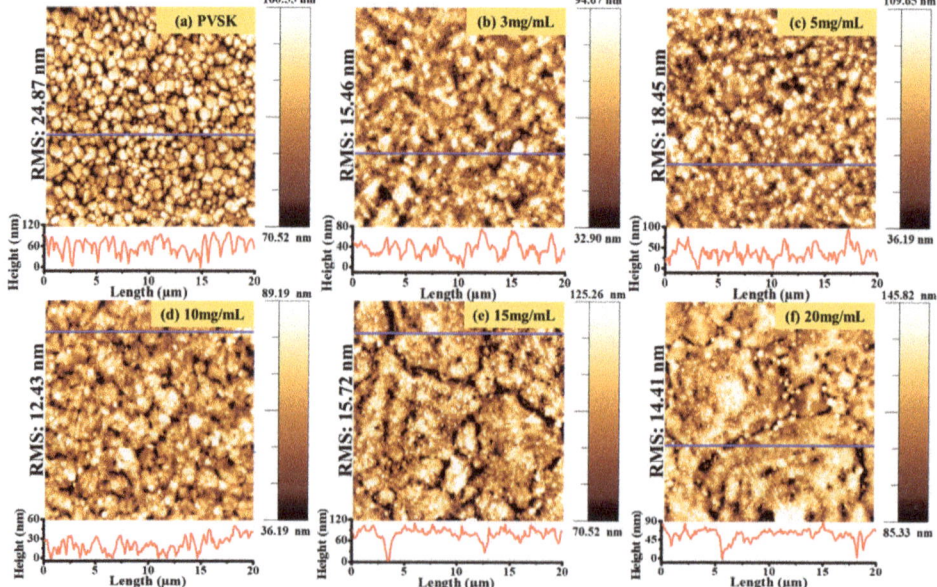

Figure 5. AFM images of perovskite films fabricated with different concentrations of NH_2-POSS. The height profiles correspond to the height variation of the blue line in the AFM images.

Figure 6 shows the device performances of perovskite films modified with a NH_2-POSS concentration of 0, 3, 5, 10, 15, and 20 mg/mL. It is well known that the most important parameters for the PSCs are open-circuit (V_{oc}), short-circuit (J_{sc}), fill factor (FF), and power conversion efficiency (PCE). As shown in Figure 6d, we found that the best PCE of the perovskite solar cells is about 17.40% when the NH_2-POSS concentration is 10 mg/mL. According to the box distribution of these parameters, we found that V_{oc} increases as the concentration of NH_2-POSS increases from 0 to 10 mg/mL, which reduces the recombination of the carrier at the interface and improves the corresponding J_{sc}. Thus, the perovskite film modified with 10 mg/mL NH_2-POSS shows the best device performance. This indicates that the changes of the surface morphology and composition of the perovskite films are closely related to the device's performance. Similarly, the PCE of the perovskite solar cell increases to 15.6% from 13.3% by introducing an appropriate POSS passivation layer between the perovskite and hole transporting layers, which is useful to significantly increase the grain size of the perovskite film and also increase the short-circuit current and open-circuit voltage [35]. Otherwise, the electronic structure and surface composition of perovskite films can be adjusted by the incorporation of $CsPbBr_3$ nanocrystals, and the photovoltaic performance of the perovskite solar cell is enhanced [36]. Thus, a good way to improve the device performance is by adjusting the surface properties of the perovskite film.

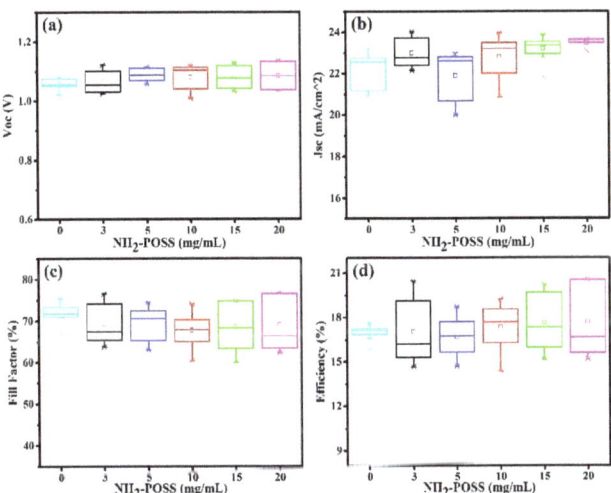

Figure 6. Photovoltaic metric for perovskite solar cells fabricated with different concentrations of NH_2-POSS. (**a**) open-circuit (V_{oc}), (**b**) short-circuit (J_{sc}), (**c**) fill factor (FF), and (**d**) power conversion efficiency (PCE).

4. Conclusions

In summary, we investigated the surface morphology, surface composition, and distribution of $FA_{0.85}MA_{0.15}Pb(I_{0.85}Br_{0.15})_3$ films after modification by different concentrations of NH_2-POSS using AFM, XPS, AR-XPS, and FTIR. It was found that the surface composition and morphology of perovskite films can be influenced by NH_2-POSS. The surface composition and its distribution can be adjusted by changing the concentration of NH_2-POSS. The surface morphology of the perovskite film was damaged when the concentration of NH_2-POSS exceeded 10 mg/mL. Thus, the proper concentration of NH_2-POSS would improve the quality of perovskite film. Our analysis provides insight for the underlining mechanisms of controlling the properties of perovskite films by adjusting the concentrations of the modified materials, which will be useful for improving the performances of perovskite solar cells.

Supplementary Materials: The following are available online at https://www.mdpi.com/article/10.3390/cryst11121544/s1, Figure S1: FTIR spectrum of perovskite with/without NH_2-POSS; Table S1: XPS fitting data for perovskite films fabricated with different concentration of NH_2-POSS.

Author Contributions: Conceptualization, H.X.; methodology, H.X. and J.L.; validation, P.Y. and J.W.; formal analysis, Y.Z. (Yuan Zhao); investigation, Y.Z. (Yangyang Zhang) and N.L.; resources, B.Y. and J.Z.; data curation, S.W.; writing—original draft preparation, Y.Z. (Yangyang Zhang); writing—review and editing, H.X., Q.C., H.H., D.N. and Y.G.; project administration, Y.G.; funding acquisition, H.X. and Y.G. All authors have read and agreed to the published version of the manuscript.

Funding: This research was funded by the National Natural Science Foundation of China (Grant Nos. 51802355), the National Key Research and Development Program of China (Grant Nos. 2017YFA0206602), the Natural Science Foundation of Hunan Province (Grant No. 2018JJ3625), the Project of High-Level Talents Accumulation of Hunan Province (Grant No. 2018RS3023) and the National Science Foundation (Grant Nos. DMR 1903962).

Institutional Review Board Statement: Not applicable.

Informed Consent Statement: Not applicable.

Data Availability Statement: Not applicable.

Acknowledgments: We thank the financial support by the National Natural Science Foundation of China (Grant Nos. 51802355) and the National Key Research and Development Program of China (Grant Nos. 2017YFA0206602). H.X. acknowledges the support by the Natural Science Foundation of Hunan Province (Grant No. 2018JJ3625). Y.G. acknowledges the support by the Project of High-Level Talents Accumulation of Hunan Province (Grant No. 2018RS3023) and the National Science Foundation (Grant Nos. DMR 1903962).

Conflicts of Interest: The authors declare no conflict of interest.

References

1. Green, M.A.; Ho-Baillie, A.; Snaith, H.J. The emergence of perovskite solar cells. *Nat. Photonics* **2014**, *8*, 506–514. [CrossRef]
2. Jeon, N.J.; Noh, J.H.; Yang, W.S.; Kim, Y.C.; Ryu, S.; Seo, J.; Seok, S.L. Compositional engineering of perovskite materials for high-performance solar cells. *Nature* **2015**, *517*, 476–480. [CrossRef]
3. Stranks, S.D.; Snaith, H.J. Metal-halide perovskites for photovoltaic and light-emitting devices. *Nat. Nanotechnol.* **2015**, *10*, 391–402. [CrossRef]
4. Huang, J.; Yuan, Y.; Shao, Y.; Yan, Y. Understanding the physical properties of hybrid perovskites for photovoltaic applications. *Nat. Rev. Mater.* **2017**, *2*, 17042. [CrossRef]
5. Xing, G.C.; Mathews, N.; Sun, S.Y.; Lim, S.S.; Lam, Y.M.; Gratzel, M.; Mhaisalkar, S.; Sum, T.C. Long-range balanced electron- and hole-transport lengths in organic-inorganic $CH_3NH_3PbI_3$. *Science* **2013**, *342*, 344–347. [CrossRef]
6. Saliba, M.; Matsui, T.; Domanski, K.; Seo, J.Y.; Ummadisingu, A.; Zakeeruddin, S.M.; Correa-Baena, J.P.; Tress, W.R.; Abate, A.; Hagfeldt, A.; et al. Incorporation of rubidium cations into perovskite solar cells improves photovoltaic performance. *Science* **2016**, *354*, 206–209. [CrossRef]
7. Gonzalez-Pedro, V.; Juarez-Perez, E.J.; Arsyad, W.S.; Barea, E.M.; Fabregat-Santiago, F.; Mora-Sero, I.; Bisquert, J. General working principles of $CH_3NH_3PbX_3$ perovskite solar cells. *Nano Lett.* **2014**, *14*, 888–893. [CrossRef] [PubMed]
8. Xiao, J.W.; Liu, L.; Zhang, D.L.; De Marco, N.; Lee, J.W.; Lin, O.; Chen, Q.; Yang, Y. The emergence of the mixed perovskites and their applications as solar cells. *Adv. Energy Mater.* **2017**, *7*, 1700491. [CrossRef]
9. NREL. Best Research-Cell Efficiencies. Available online: https://www.nrel.gov/pv/assets/pdfs/best-research-cell-efficiencies-rev211117.pdf (accessed on 17 November 2021).
10. Conings, B.; Baeten, L.; De Dobbelaere, C.; D'Haen, J.; Manca, J.; Boyen, H.G. Perovskite-based hybrid solar cells exceeding 10% efficiency with high reproducibility using a thin film sandwich approach. *Adv. Mater.* **2014**, *26*, 2041–2046. [CrossRef] [PubMed]
11. Geissbuhler, J.; Werner, J.; de Nicolas, S.M.; Barraud, L.; Hessler-Wyser, A.; Despeisse, M.; Nicolay, S.; Tomasi, A.; Niesen, B.; De Wolf, S.; et al. 22.5% efficient silicon heterojunction solar cell with molybdenum oxide hole collector. *Appl. Phys. Lett.* **2015**, *107*, 081601. [CrossRef]
12. Zhou, H.P.; Chen, Q.; Li, G.; Luo, S.; Song, T.B.; Duan, H.S.; Hong, Z.R.; You, J.B.; Liu, Y.S.; Yang, Y. Interface engineering of highly efficient perovskite solar cells. *Science* **2014**, *345*, 542–546. [CrossRef]
13. Sanehira, E.M.; de Villers, B.J.T.; Schulz, P.; Reese, M.O.; Ferrere, S.; Zhu, K.; Lin, L.Y.; Berry, J.J.; Luther, J.M. Influence of electrode interfaces on the stability of perovskite solar cells: Reduced degradation using MoOx/Al for hole collection. *ACS Energy Lett.* **2016**, *1*, 38–45. [CrossRef]
14. Chen, B.; Rudd, P.N.; Yang, S.; Yuan, Y.B.; Huang, J.S. Imperfections and their passivation in halide perovskite solar cells. *Chem. Soc. Rev.* **2019**, *48*, 3842–3867. [CrossRef]

15. Steirer, K.X.; Schulz, P.; Teeter, G.; Stevanovic, V.; Yang, M.; Zhu, K.; Berry, J.J. Defect tolerance in methylammonium lead triiodide perovskite. *ACS Energy Lett.* **2016**, *1*, 360–366. [CrossRef]
16. Li, N.X.; Tao, S.X.; Chen, Y.H.; Niu, X.X.; Onwudinanti, C.K.; Hu, C.; Qiu, Z.W.; Xu, Z.Q.; Zheng, G.H.J.; Wang, L.G.; et al. Cation and anion immobilization through chemical bonding enhancement with fluorides for stable halide perovskite solar cells. *Nat. Energy* **2019**, *4*, 408–415. [CrossRef]
17. Zhu, X.J.; Yang, D.; Yang, R.X.; Yang, B.; Yang, Z.; Ren, X.D.; Zhang, J.; Niu, J.Z.; Feng, J.S.; Liu, S.Z. Superior stability for perovskite solar cells with 20% efficiency using vacuum co-evaporation. *Nanoscale* **2017**, *9*, 12316–12323. [CrossRef] [PubMed]
18. Yang, W.S.; Park, B.W.; Jung, E.H.; Jeon, N.J.; Kim, Y.C.; Lee, D.U.; Shin, S.S.; Seo, J.; Kim, E.K.; Noh, J.H.; et al. Iodide management in formamidinium-lead-halide–based perovskite layers for efficient solar cells. *Science* **2017**, *356*, 1376–1379. [CrossRef] [PubMed]
19. De Marco, N.; Zhou, H.P.; Chen, Q.; Sun, P.Y.; Liu, Z.H.; Meng, L.; Yao, E.P.; Liu, Y.S.; Schiffer, A.; Yang, Y. Guanidinium: A route to enhanced carrier lifetime and open-circuit voltage in hybrid perovskite solar cells. *Nano Lett.* **2016**, *16*, 1009–1016. [CrossRef]
20. Noel, N.K.; Abate, A.; Stranks, S.D.; Parrott, E.S.; Burlakov, V.M.; Goriely, A.; Snaith, H.J. Enhanced photoluminescence and solar cell performance via Lewis base passivation of organic-inorganic lead halide perovskites. *ACS Nano* **2014**, *8*, 9815–9821. [CrossRef] [PubMed]
21. Son, D.Y.; Kim, S.G.; Seo, J.Y.; Lee, S.H.; Shin, H.; Lee, D.; Park, N.G. Universal approach toward hysteresis-free perovskite solar cell via defect engineering. *J. Am. Chem. Soc.* **2018**, *140*, 1358–1364. [CrossRef] [PubMed]
22. Liu, S.J.; Guo, R.Q.; Li, C.; Lu, C.F.; Yang, G.C.; Wang, F.Y.; Nie, J.Q.; Ma, C.; Gao, M. POSS hybrid hydrogels: A brief review of synthesis, properties and applications. *Eur. Polym. J.* **2021**, *143*, 110180. [CrossRef]
23. Peng, B.; Liu, R.; Han, G.; Wang, W.; Zhang, W.Q. The synthesis of thermoresponsive POSS-based eight-arm star poly (N-isopropylacrylamide): Comparison between Z-RAFT and R-RAFT strategies. *Polym. Chem.* **2021**, *12*, 2063–2074. [CrossRef]
24. Zhao, B.J.; Mei, H.G.; Liu, N.; Zheng, S.X. Organic–inorganic polycyclooctadienes with double-decker silsesquioxanes in the main chains: Synthesis, self-healing, and shape memory properties regulated with quadruple hydrogen bonds. *Macromolecules* **2020**, *53*, 7119–7131. [CrossRef]
25. Li, G.Z.; Wang, L.C.; Ni, H.L.; Pittman, C.U. Polyhedral oligomeric silsesquioxane (POSS) polymers and copolymers: A review. *J. Inorg. Organomet. Polym. Mater.* **2001**, *11*, 123–154. [CrossRef]
26. Huang, H.; Lin, H.; Kershaw, S.V.; Susha, A.S.; Choy, W.C.H.; Rogach, A.L. Polyhedral oligomeric silsesquioxane enhances the brightness of perovskite nanocrystal-based green light-emitting devices. *J. Phys. Chem. Lett.* **2016**, *7*, 4398–4404. [CrossRef] [PubMed]
27. Liu, N.; Du, Q.; Yin, G.Z.; Liu, P.F.; Li, L.; Xie, H.P.; Zhu, C.; Li, Y.J.; Zhou, H.P.; Zhang, W.B.; et al. Extremely low trap-state energy level perovskite solar cells passivated using NH_2-POSS with improved efficiency and stability. *J. Mater. Chem. A* **2018**, *6*, 6806–6814. [CrossRef]
28. Xie, H.P.; Niu, D.M.; Lyu, L.; Zhang, H.; Zhang, Y.H.; Liu, P.; Wang, P.; Wu, D.; Gao, Y.L. Evolution of the electronic structure of C_{60}/$La_{0.67}Sr_{0.33}MnO_3$ interface. *Appl. Phys. Lett.* **2016**, *108*, 011603. [CrossRef]
29. Xie, H.P.; Huang, H.; Cao, N.T.; Zhou, C.H.; Niu, D.M.; Gao, Y.L. Effects of annealing on structure and composition of LSMO thin films. *Phys. B Condens. Matter* **2015**, *477*, 14–19. [CrossRef]
30. Liu, P.; Liu, X.L.; Lyu, L.; Xie, H.P.; Zhang, H.; Niu, D.M.; Huang, H.; Bi, C.; Xiao, Z.G.; Huang, J.S.; et al. Interfacial electronic structure at the $CH_3NH_3PbI_3$/MoO_x interface. *Appl. Phys. Lett.* **2015**, *106*, 506–902. [CrossRef]
31. Zhang, L.; Yang, Y.G.; Huang, H.; Lyu, L.; Zhang, H.; Cao, N.T.; Xie, H.P.; Gao, X.Y.; Niu, D.M.; Gao, Y.L. Thickness-dependent air-exposure-induced phase transition of CuPc ultrathin films to well-ordered one-dimensional nanocrystals on layered substrates. *J. Phys. Chem. C* **2015**, *119*, 4217–4223. [CrossRef]
32. Gao, D.G.; Wang, P.P.; Shi, J.B.; Li, F.; Li, W.B.; Lyu, B.; Ma, J.Z. A green chemistry approach to leather tanning process: Cage-like octa (aminosilsesquioxane) combined with Tetrakis (hydroxymethyl) phosphonium sulfate. *J. Clean. Prod.* **2019**, *229*, 1102–1111. [CrossRef]
33. Xie, H.P.; Liu, X.L.; Lyu, L.; Niu, D.M.; Wang, Q.; Huang, J.S.; Gao, Y.L. Effects of precursor ratios and annealing on electronic structure and surface composition of $CH_3NH_3PbI_3$ perovskite films. *J. Phys. Chem.* **2015**, *120*, 215–220. [CrossRef]
34. Feng, W.; Geng, W.; Zhou, Y.; Fang, H.H.; Tong, C.J.; Loi, M.A.; Liu, L.M.; Zhao, N. Phenylalkylamine passivation of organolead halide perovskites enabling high-efficiency and air-stable photovoltaic cells. *Adv. Mater.* **2016**, *28*, 9986–9992.
35. Liu, B.T.; Lin, H.R.; Lee, R.H.; Gorji, N.E.; Chou, J.C. Fabrication and characterization of an efficient inverted perovskite solar cells with POSS passivating hole transport layer. *Nanomaterials* **2021**, *11*, 974. [CrossRef]
36. Liu, Y.Q.; Zai, H.C.; Xie, H.P.; Liu, B.X.; Wang, S.T.; Zhao, Y.; Niu, D.M.; Huang, H.; Chen, Q.; Gao, Y.L. Effects of $CsPbBr_3$ nanocrystals concentration on electronic structure and surface composition of perovskite films. *Org. Electron.* **2019**, *73*, 327–331. [CrossRef]

Article

Synthesis, Molecular Structure, Thermal and Spectroscopic Analysis of a Novel Bromochalcone Derivative with Larvicidal Activity

Pollyana P. Firmino [1,2], Jaqueline E. Queiroz [3], Lucas D. Dias [2], Patricia R. S. Wenceslau [1], Larissa M. de Souza [2], Ievgeniia Iermak [2], Wesley F. Vaz [1], Jean M. F. Custódio [4], Allen G. Oliver [4], Gilberto L. B. de Aquino [1,3] and Hamilton B. Napolitano [1,*]

1. Grupo de Química Teórica e Estrutural de Anápolis, Universidade Estadual de Goiás, Anápolis 75132-903, GO, Brazil; polly.firmino@outlook.com (P.P.F.); patricia.wenceslaubio@gmail.com (P.R.S.W.); wesfonseca@gmail.com (W.F.V.); gilbaqui@hotmail.com (G.L.B.d.A.)
2. São Carlos Institute of Physics, University of São Paulo, São Carlos 13566-590, SP, Brazil; lucasdanillodias@gmail.com (L.D.D.); larissamarila@hotmail.com (L.M.d.S.); ievgeniia.iermak@gmail.com (I.I.)
3. Laboratório de Pesquisa em Bioprodutos e Síntese, Universidade Estadual de Goiás, Anápolis 75132-903, GO, Brazil; jaqueevan.je@gmail.com
4. Department of Chemistry and Biochemistry, University of Notre Dame, Notre Dame, IN 46556, USA; jeanmfcustodio@gmail.com (J.M.F.C.); aoliver2@nd.edu (A.G.O.)
* Correspondence: hamilton@ueg.br

Abstract: Chalcones belong to the flavonoids family and are natural compounds which show promising larvicidal property against *Aedes aegypti* larvae. Aiming to obtain a synthetic chalcone derivative with high larvicidal activity, herein, a bromochalcone derivative, namely (*E*)-3-(4-butylphenyl)-1-(4-bromophenyl)-prop-2-en-1-one (BBP), was designed, synthesized and extensively characterized by ^{1}H- and ^{13}C- nuclear magnetic resonance (NMR), infrared (IR), Raman spectroscopy, mass spectrometry (MS), ultraviolet–visible spectroscopy (UV-Vis), thermogravimetric analysis (TGA), differential scanning calorimetry (DSC), and X-ray diffraction. Further, the quantum mechanics calculations implemented at the B3LYP/6–311+G(d)* level of the theory indicate that the supramolecular arrangement was stabilized by C–H$\cdots$O and edge-to-face C–H$\cdots\pi$ interactions. The EGAP calculated (3.97 eV) indicates a good reactivity value compared with other similar chalcone derivatives. Furthermore, the synthesized bromochalcone derivative shows promising larvicidal activity (mortality up to 80% at 57.6 mg·L^{-1}) against *Ae. aegypti* larvae.

Keywords: bromochalcone; X-ray diffraction; B3LYP/6-311+G(d); larvicide; *A. aegypti* larvae

Citation: Firmino, P.P.; Queiroz, J.E.; Dias, L.D.; Wenceslau, P.R.S.; de Souza, L.M.; Iermak, I.; Vaz, W.F.; Custódio, J.M.F.; Oliver, A.G.; de Aquino, G.L.B.; et al. Synthesis, Molecular Structure, Thermal and Spectroscopic Analysis of a Novel Bromochalcone Derivative with Larvicidal Activity. *Crystals* **2022**, *12*, 440. https://doi.org/10.3390/cryst12040440

Academic Editor: Lilianna Checinska

Received: 1 March 2022
Accepted: 18 March 2022
Published: 22 March 2022

Publisher's Note: MDPI stays neutral with regard to jurisdictional claims in published maps and institutional affiliations.

Copyright: © 2022 by the authors. Licensee MDPI, Basel, Switzerland. This article is an open access article distributed under the terms and conditions of the Creative Commons Attribution (CC BY) license (https://creativecommons.org/licenses/by/4.0/).

1. Introduction

Chalcones are open-chain flavonoid compounds with an α,β-unsaturated, carbonyl group consisting of two aromatic rings with a range of substituents [1,2]. These molecules can be isolated from pigments commonly found in plants or synthesized by the classic Claisen–Schmidt method [3–5]. This class is a privileged structure with biological activities such as anti-inflammatory [6], anticancer [7,8], antifungal [9,10], antibacterial [2], antioxidant [11], and antiparasitary [12,13]. Likewise, substituting groups on the aromatic rings of chalcone can modulate its larvicidal activity. In this context, the understanding of the structure–activity relationship of chalcones and their larvicidal activity is still a challenge.

Yellow fever, chikungunya, dengue fever, and Zika are infectious diseases transmitted by the bite of the *Aedes aegypti* mosquito [14]. It is calculated that more than 2.5 billion people live in transmission risk areas (e.g., Brazil, some African countries) [15]. Due to the absence of a prophylactic vaccine, these infectious diseases can only be contained by combating the vector insect [16]. Nowadays, one of the most accepted methods for controlling infectious diseases transmitted by *Ae. aegypti* is through the application of chemical larvicides. Organophosphate, organochlorine, and synthetic pyrethroid insecticides are used in public health control measures [17]. Nevertheless, these larvicides still show some issues e.g., (i) low selectivity, (ii) environment damage, and (iii) toxicity for humans. In this regard, chalcones (a natural product) have been tested as a potential larvicide, showing promising results [18–21].

To obtain a chalcone derivative with larvicidal activity, herein, we reported the design, synthesis, and structural description of a chalcone, namely (E)-3-(4-butylphenyl)-1-(4-bromophenyl)-prop-2-en-1-one (BBP) (Figure 1), using the density functional theory (DFT) implemented at the B3LYP/6–311+G(d)* level of the theory [22–24]. Moreover, this work describes the full characterization of BBP by nuclear magnetic resonance (NMR), infrared (IR), Raman spectroscopy, mass spectrometry (MS), ultraviolet–visible spectroscopy (UV-Vis), thermogravimetric analysis (TG), and differential scanning calorimetry (DSC). Additionally, the supramolecular arrangements were analyzed by X-ray diffraction and Hirshfeld surfaces (HS) to understand the electrophilic and nucleophilic reactions, and hydrogen bond interactions' connectivity [22,25] was observed, which may influence its larvicidal activity. Furthermore, the larvicidal activity (mortality) of BBP against *Ae. aegypti* larvae (L3) were evaluated.

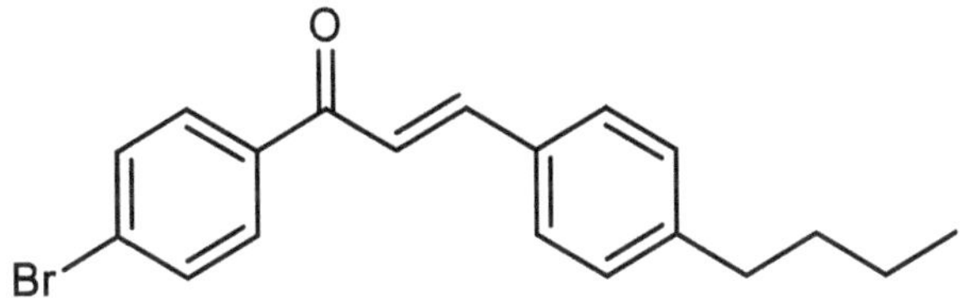

Figure 1. Molecular structure of 3-(4-butylphenyl)-1-(4-bromophenyl)-prop-2-en-1-one (BBP).

2. Materials and Methods

2.1. Synthesis and Crystallization

Chemicals and solvents required for BBP synthesis were obtained from commercial sources and used without further purification. 4-Butylbenzaldehyde (1 mmol; 162.3 mg) and 4'-bromo-acetophenone (1 mmol; 199.0 mg) were added in ethanol (0.5 mL). Then, pulverized KOH (1 mmol; 56.1 mg) was added, and the reaction mixture was kept under manual shaking for 3 min at 25 °C. The reaction's progress was monitored by thin-layer chromatography (TLC) (silica gel 60 UV254 plate), using CH_2Cl_2 as an eluent. Then, crystals were obtained and collected by vacuum filtration, followed by a crystallization process using ethanol as a solvent. BBP was obtained in 95% yield (0.95 mmol; 326.0 mg). $C_{19}H_{19}BrO$ (342.06 g/mol); white solid, and m.p. 109.8 °C. To grow BBP crystals, the finely powdered sample was added to a conical flask with a known volume of dichloromethane, then the solution was recrystallized by the diffusion of pentane vapor in a dichloromethane solution and kept at room temperature for slow evaporation for 72 h until crystal formation.

2.2. Spectroscopic and Thermal Characterization

A Bruker Avance 500MHZ NMR spectrometer was used to obtain ^{1}H- and ^{13}C-NMR spectra using TMS as the internal standard and $CDCl_3$ as the solvent. The corresponding shifts reveal the purity of the compound.

The IR spectrum was recorded using a Perkin Elmer (8400S FT-IR) spectrometer in the wavenumber range 400–4000 cm^{-1} by the KBr technique. Raman measurements were performed on a WITec Alpha 300 RAS microscope (WITec, Ulm, Germany). The excitation wavelength was 785 nm, the detection range was 100–4000 cm^{-1}, with the resolution of 1 cm^{-1}. The spectra were collected with a 20× magnification objective (Zeiss, Jena, Germany). Spectra were recorded with the integration time of 30 s and 2 accumulations. Obtained spectra were processed using WITec Project FOUR and Origin 2016 software.

Gas chromatogram and mass spectrum (GC-MS) were recorded in Shimadzu (QP2010 Ultra) equipped with capillary column CBP-5 (30 × 0.25 × 0.25); the injection volume was 1.0 µL in split mode and helium as drag gas with 1.0 mL.min-1 flow. The injector temperature was 280 °C, and the detector was 310 °C. The initial oven temperature was 100 °C (for 2 min), followed by a heating ramp of 30 °C.min^{-1} till 300 °C, then 300 °C for 10 min. Solid-state UV-vis spectra were obtained using a Cary 5000 UV-Vis-NIR spectrometer. IR (KBr): 1659 υ(C=O); 1599 υ(C=C); 1419 υ(C=C) aromatic ring; 987 υ(CH=CH); 571 υ(C–Br) cm^{-1}; FT-Raman: 1659 υ(C=O); 1599 υ(C=C); 1421 υ(C=C) aromatic ring; 989 υ(CH=CH); 699 υ(C–Br) cm^{-1}; MS (m/z): 342, 344, 287, 285, 263 and 57; UV-vis (DMSO): λ = 213, 336 nm.

TGA was performed in the Perkin Elmer Pyris (1TGA model) equipment. The evaluation was carried out with a 2.131 mg sample mass in an alumina crucible. The TGA was performed on the BBP compound at an initial temperature ranging from 25 °C to 600 °C, with a heating rate of 10 °C·min^{-1} under nitrogen purge gases with 20 mL·min^{-1} flow. DSC (214 Polyma, Netzsch) measurements were carried out in platinum crucibles with the crucibles sealed without drilling the lid. A sample weighing 1.91 mg was used in the crucibles. The temperature range was 20–200 °C applied at a heating rate of 10 °C·min^{-1}, under a dynamic 5.0 N$_2$ atmosphere of 40 mL·min^{-1}. For these analyses, a correction was performed to eliminate the background.

2.3. X-ray Diffraction Analysis

A suitable crystal of BPP was selected, kept at 119.99 K, and the X-ray diffraction was collected on a Bruker APEX-II CCD diffractometer with wavelength radiation MoKα λ = 0.71073 Å. The size of the crystal used was 0.226 × 0.197 × 0.044 mm^3. The structure was solved by direct methods with the intrinsic phasing method (SHELXT) [26] using Olex2 1.3 version [27] and refined by least-squares minimization with SHELXL [28]. All hydrogen atoms were fixed at the calculated position, and potential hydrogen–bond interactions were verified through Platon [29]. The structure data were deposited in the Cambridge Crystallography Data Center (CCDC) under code 2069836.

The HS was generated using a standard (high) surface resolution with the three-dimensional dnorm surfaces mapped over a red to blue [30] fixed-color scale, which is shorter than the white surface that indicates contacts with distances equal to the sum of van der Waals [31]. HS is a tool to assess observed intermolecular interactions and is defined by the weight function w(r) [32]. It is calculated for each atom in a crystal by Equation (1) and can be determined using Crystal Explorer software [33,34].

$$w(r) = \frac{\sum_{molecule} \rho_i^{at}(r)}{\sum_{crystal} \rho_i^{at}(r)} \quad (1)$$

The ρ(r) is the average density, the numerator is a sum over the atoms in the molecule, and the denominator is the sum over the crystal [33].

2.4. Theoretical Analysis

Electronic calculations were undertaken from experimentally determined atom coordinates. The geometry optimization and vibrational modes were carried out using the Gaussian 09 [35] packages, through DFT applying B3LYP/6-311+G(d)* as the functional and basis set to confirm the stability associated with optimized geometries of molecules [36–38], and a scale factor of 0.9686 was used to mitigate the effects of overestimation in the vibration values calculated via DFT, obtaining better convergence with the experimental values [39–42]. To check that the optimized geometry was found in a local minimum, analytic harmonic frequency calculations were undertaken at the identical level of theory. The absence of imaginary frequencies reveals that the optimized structure is truly at a local minimum. The highest occupied molecular orbital (HOMO), the lowest unoccupied molecular orbital (LUMO), and the molecular electrostatic potential (MEP) map were graphically represented by Gaussview 6.0 software (Wallingford, Connecticute, USA) [43].

2.5. Larvicidal Assays

Aedes aegypti (Rockefeller strain) eggs were provided by Prof. Alex Martins Machado from the Laboratory of Virology and Cell Culture at the Federal University of Mato Grosso do Sul, MS, Brazil. These eggs were hatched in a plastic tray containing 1.5 L of dechlorinated water and kept in an incubator under controlled conditions (at 27 ± 2 °C under 12:12 h light/dark photoperiod with $73.0 \pm 0.4\%$ relative humidity). Once a day, the larvae were fed with a blend of powdered fish food AlconBASIC® MEP 200 Complex (Alcon, Camboriú, SC, Brazil) and dry yeast, in a 3:1 proportion. Larvae were collected at the third stage (L3) to be used in the larvicidal assays.

Larvicidal activity assays were performed according to Targanski et al. (2021) [18]. Each bioassay consisted of incubating 10 larvae (L3) in distilled water (20 mL), containing different concentrations of chalcone (7.2–57.6 mg·L^{-1}). An equivalent volume to the highest concentration tested (57.6 mg·L^{-1}) was used as a negative control. Larval mortality (%) was recorded after 48 h. All experiments were carried out in triplicate. The LC25, LC50, and LC75 values of *Ae. aegypti* larvae were estimated by non-linear regression, calculated using Origin 2020 software (OriginLab Corp. 9.7).

3. Results

3.1. Synthesis and Crystallization

BBP chalcone was synthesized by the Claisen–Schmidt condensation protocol. In this procedure, 4-butylbenzaldehyde, 4′-bromo-acetophenone, and KOH were added in ethanol and were reacted by manual shaking for 3 min, at 25 °C (Scheme 1). BBP was obtained in 95% yield (0.95 mmol; 326.0 mg). High-quality crystals of BBP were obtained, and the molecular modeling analysis was performed.

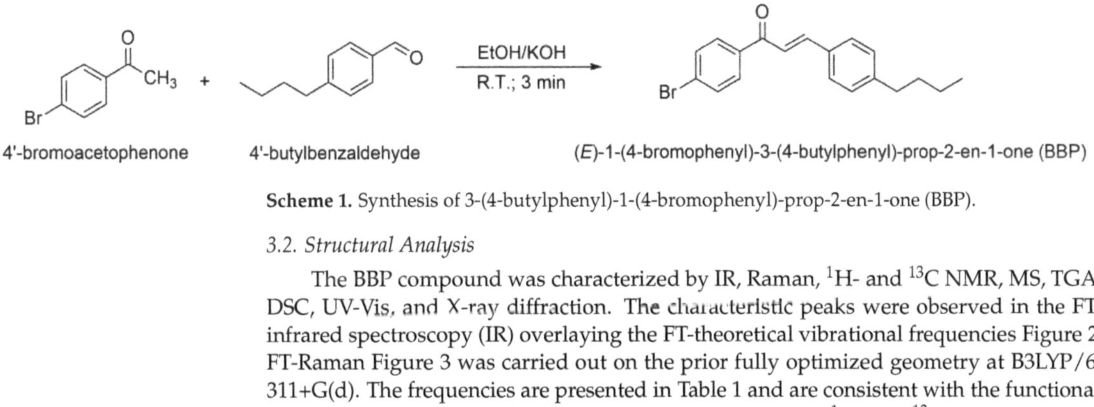

Scheme 1. Synthesis of 3-(4-butylphenyl)-1-(4-bromophenyl)-prop-2-en-1-one (BBP).

3.2. Structural Analysis

The BBP compound was characterized by IR, Raman, ^{1}H- and ^{13}C NMR, MS, TGA, DSC, UV-Vis, and X-ray diffraction. The characteristic peaks were observed in the FT-infrared spectroscopy (IR) overlaying the FT-theoretical vibrational frequencies Figure 2. FT-Raman Figure 3 was carried out on the prior fully optimized geometry at B3LYP/6-311+G(d). The frequencies are presented in Table 1 and are consistent with the functional groups present in the BBP. Moreover, Figures 4 and 5 show ^{1}H- and ^{13}C-NMR spectra of BBP, respectively. These spectra were used to identify the sample purity and structural conformation of BBP. From the analysis of the spectra, the following chemical shifts are present: ^{1}H-NMR (500 MHz, CDCl$_3$) δ 7.87 (d, J = 8.6 Hz, 1H), 7.79 (d, J = 15.7 Hz, 1H), 7.63 (d, J = 8.6 Hz, 1H), 7.55 (d, J = 8.1 Hz, 1H), 7.42 (d, J = 15.6 Hz, 1H), 7.23 (d, J = 8.1 Hz, 1H), 2.65 (d, J = 7.7 Hz, 1H), 1.62 (dq, J = 12.9, 7.6 Hz, 1H), 1.43–1.31 (m, 1H), 0.93 (t, J = 7.4 Hz, 2H); ^{13}C-NMR (126 MHz, CDCl$_3$): δ 195.37, 144.42, 136.81, 133.86, 133.52, 132.27, 132.24, 131.88, 130.46, 129.86, 128.52, 127.60, 127.49, 127.30, 126.58, 125.60, 125.04, 124.53.

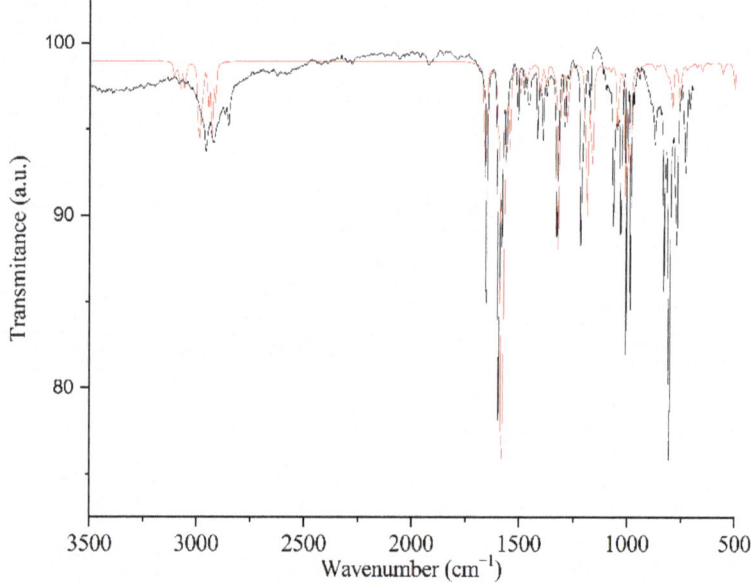

Figure 2. IR Overlay for FT-infrared spectrum of BBP in black and FT-theoretical vibrational frequencies in red.

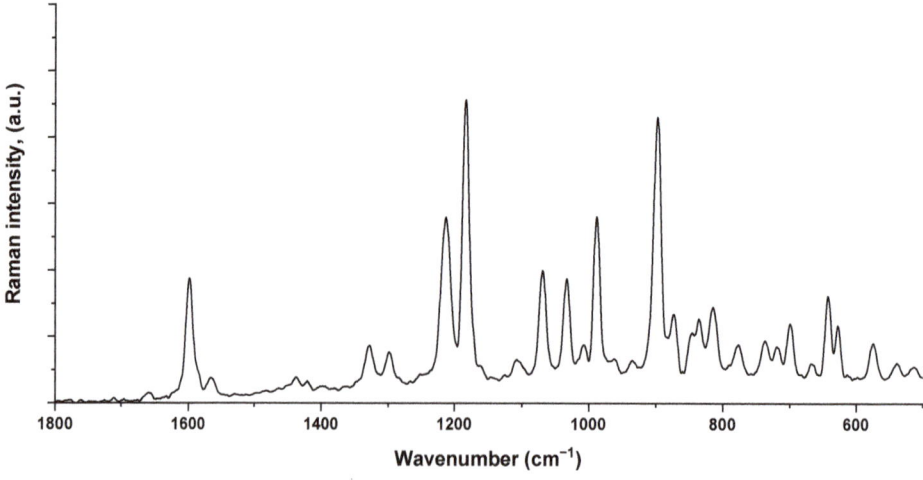

Figure 3. FT-Raman spectrum of BBP.

Table 1. Assignments of FT-IR; theoretical FT-IR and FT-Raman spectral bands for BBP.

Infrared Bands (cm^{-1})	Theoretical Infrared Bands (cm^{-1})	Raman Bands (cm^{-1})	Assignment
2953	3051.14–3107.11	2993	υ(C–H) aromatic
2924	2914.51–2992.48	–	υ(C–H)
2853	3051.14	–	υ(=C–H)
1659	1662.49	1661	υ(C=O)
1599	1582.93	1599	υ(C=C)
1419	1547.44–1603.37	1421	υ(C=C) aromatic ring
1330	1302.72	1328	δ(–C–H)
–	–	1299	δ(C–H)
1219	–	1297, 1327, 1421, 1437	δ(C–C)
1185	1113.46–1172.86	1180	δ(C–H) aromatic in plane
1068	1039.37	1069	υ(C–C) alicyclic, aliphatic chain vibrations
987	990.73	989	υ(CH=CH)
832, 810, 773	816.92	898	δ(C=O) in plane
571	–	699	υ(C–Br)
–		703, 716, 738, 776, 813, 836, 871, 898, 990	δ(C–C) aliphatic chains

υ: stretching vibrations. δ: bending vibrations.

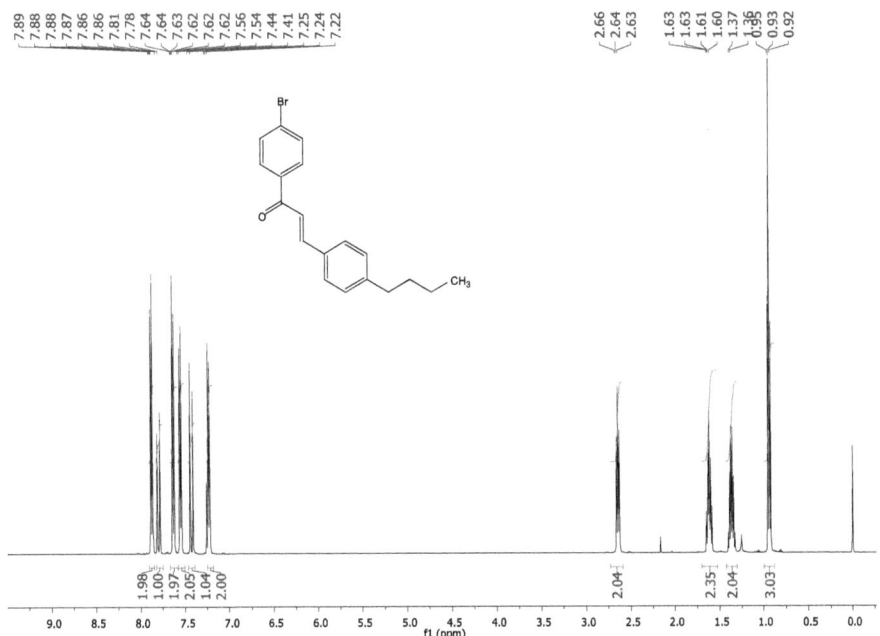

Figure 4. ^{1}H-NMR spectrum of BBP.

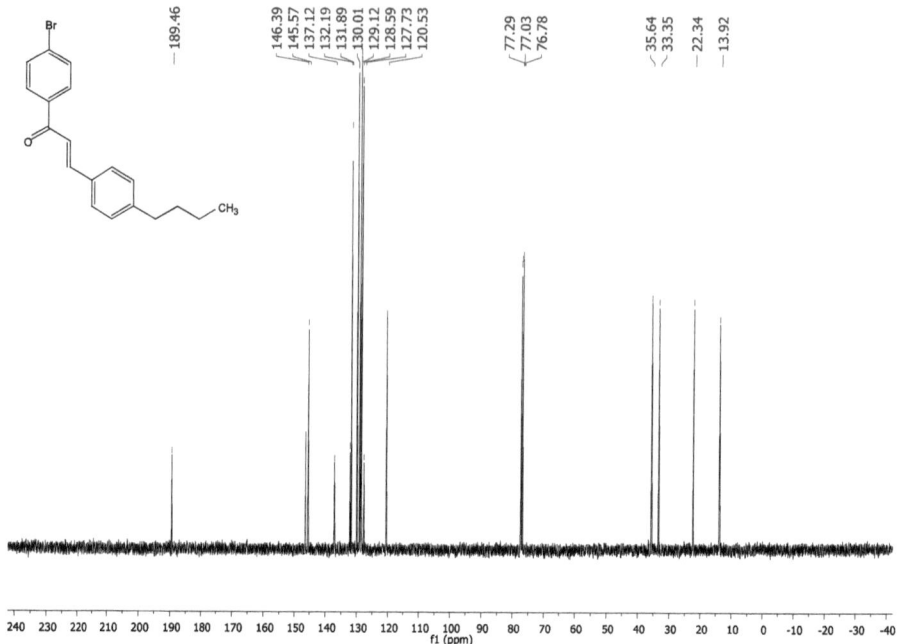

Figure 5. ^{13}C-NMR spectrum of BBP.

The results of the theoretical FT-IR show that the aromatic C–H stretching band occurs in the range of 3052–3107 cm^{-1}; experimentally, this vibration is observed at 2953 cm^{-1}. The C–H stretching vibrations emerge at the range of 2914–2992 cm^{-1}; experimentally, this vibration is observed at 2924 cm^{-1}. Regarding the C–H stretching vibrations in the vinyl group, these were observed at 3051 cm^{-1}, while experimentally they were observed at 2853 cm^{-e}. The theoretical bands observed at 1662 cm^{-1} are assigned to stretching of the C=O group, while the experimental band is observed at 1659 cm^{-1}. In terms of C=C stretching, this is assigned a calculated band at 1582 cm^{-1}; this value is closer to that obtained experimentally: 1599 cm^{-1}. Furthermore, the calculated –C–H bending vibration modes were found at 1302 cm^{-1}, while experimentally they arise at 1330 cm^{-1}. Moreover, for aromatic C–H in plane mode, the theoretical value is 1113–1172 cm^{-1}; experimentally, it occurs at 1185 cm-1. The alicyclic C–C and CH=CH theoretical stretching vibrations are observed at 1039 and 990 cm^{-1}, respectively. Meanwhile, experimental bands are assigned at 1068 and 987 cm^{-1}, respectively. Finally, the C=O in-plane bending vibrations arise at 816 cm^{-1}, while the experimental values are in the 832, 810, and 773 cm^{-1} regions. All these values are listed in Table 1.

The theoretical IR band data were supported by analyses from the animation option of Gaussview 6 [43]. The observed differences between the theoretical and the experimental findings could be explained by the fact that experiments were performed on a crystalline conformation of the sample, while the theoretical calculations were made for a free molecule, thus being systematically overestimated for vibrational frequencies [22]. Raman spectroscopy was also used to characterize and determine the quality/purity of the crystals of BBP. Spectra for each crystal type were normalized to a certain peak (the exact position of the peak is indicated by arrows on the graphs and in the figure legends). For the Raman spectrum, the C–H stretching vibrations were observed at 1300–1000 cm^{-1}, showing a strong Raman intensity. The C–H out-of-plane deformations were observed at 703, 716, 738, 776, 813, 836, 871, 898, and 990 cm^{-1}. Moreover, peaks at 1297, 1327, 1421, and 1437 cm^{-1} crystal were observed and assigned to C–C stretching vibrations. A sharp band at 1599 cm^{-1} was also observed and assigned to ethylenic bridge vibrations. The C=O vibrations were also observed as a weak Raman band at 1661 cm^{-1}. The C=O stretching vibration is further influenced by the intermolecular hydrogen bonding between the C=O group and phenyl ring, which may explain the weak Raman band observed. A Raman peak at 1070 cm^{-1} was used to normalize the spectrum. The characteristic curve and peaks observed in the TGA and DTG, respectively, are shown in Figure 6.

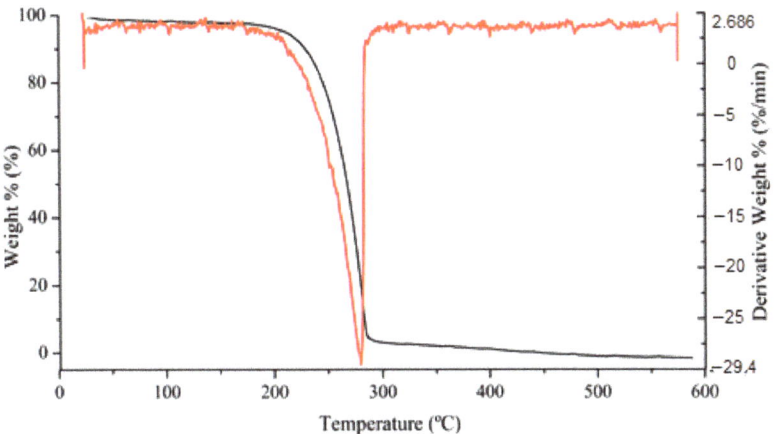

Figure 6. TGA and DTG of BBP.

Through the TGA curve, the decomposition of the BBP compound was observed as a well-defined event, represented by a single DTG peak. The DTG allowed us to see the temperatures corresponding to the beginning and end of the thermal event. The compound maintained its mass at temperatures close to 210 °C; then, after this point, one interval of great mass loss occurred until 300 °C, reaching almost 0% of the weight. Figure 7 displays the DSC traces and indicates a sharp exothermic peak at 109.8 °C for the heating of the chalcone crystal. The DSC thermogram indicates direct melting of the crystal phase to the isotropic liquid phase since the thermogram showed a single endotherm during the heating cycles. The enthalpy change (ΔH) was 31.71 kJ.mol^{-1}.

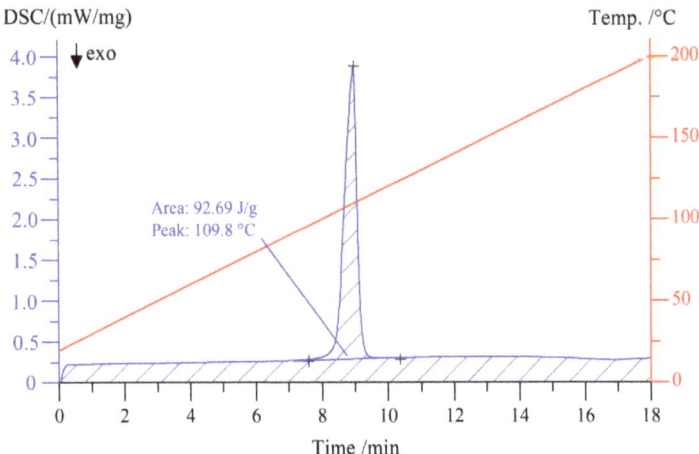

Figure 7. DSC measurement of BBP.

The UV-Vis absorption spectrum of BBP, Figure 8, clearly shows an absorption band at 336 nm, which can be attributed to the functional group —C=C—CO and phenyl ring chromophore and shows a similarity with the chalcone-type molecules.

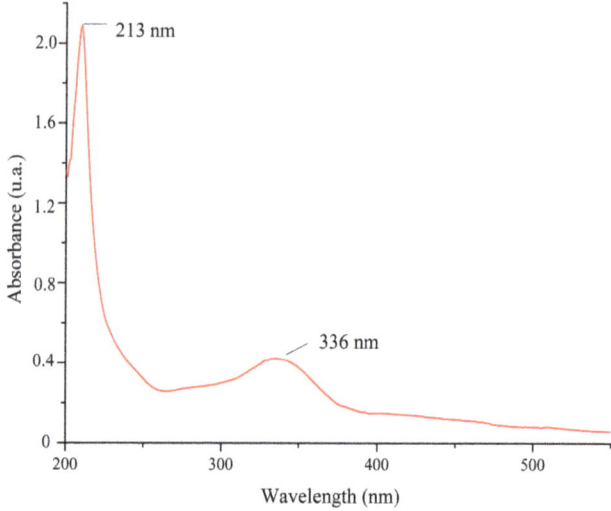

Figure 8. UV-vis spectrum of BBP in DMSO.

BBP shows a butyl group and a bromine atom ortho-bonded to aromatic rings A and B, respectively. It crystallizes into the Pbca space group with a single molecule in the asymmetric unit. The ortep representation, Figure 9, shows atomic ellipsoids with similar size and absence of disorder problem [39]. Comprehensive statistical analyses by Mogul software [44] confirmed the absence of any unusual bond distances, bond angles, and torsions on observed conformation. The main crystallographic refinement data are presented in Table 2.

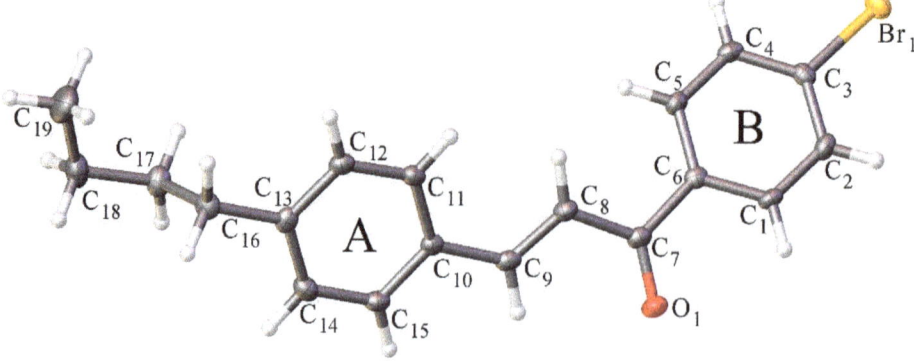

Figure 9. Ortep diagram showing BBP molecule in the asymmetric unit drawing with 50% probability ellipsoid.

Table 2. Crystallographic parameters for BBP.

	Crystal Data
Chemical formula	$C_{19}H_{19}BrO$
Mr	343.25
Crystal system, space group	Orthorhombic, Pbca
Temperature (K)	120
a, b, c (Å)	11.3065 (12), 8.1831 (9), 33.892 (4)
V (Å3)	3135.8 (6)
Z	8
Radiation type	Mo Kα
μ (mm^{-1})	2.62
Crystal size (mm)	0.23 × 0.20 × 0.04
	Data Collection
Diffractometer	Bruker APEX-II CCD
Absorption correction	Multi-scan SADABS2016/2 (Bruker,2016/2) was used for absorption correction. wR2(int) was 0.0455 before and 0.0414 after correction. The Ratio of minimum to maximum transmission is 0.7126. The λ/2 correction factor is not present.
Tmin, Tmax	0.662, 0.929
No. of measured, independent and observed [I > 2σ(I)] reflections	42,969, 3878, 2966
Rint	0.062
(sin θ/λ)max (Å^{-1})	0.666
	Refinement
R [F2 > 2σ(F2)], wR(F2), S	0.032, 0.068, 1.03
No. of reflections	3878
No. of parameters	191
H-atom treatment	H-atom parameters constrained
Δρmax, Δρmin (e Å^{-3})	0.39, −0.39

Computer programs: Olex2.solve 1.3 [45] SHELXL 2018/3 [46], Olex2 1.3 [27].

Molecular planarity is an important parameter; the higher the number of donor–receptor interactions in a molecule, the better the electron transfer and nonlinearity. The planarity in the chalcones is related to donor–acceptor interactions across the main chain of chalcones, to π-conjugation throughout charge transfer [47,48]. The angle between the aromatic rings from BBP is 47.05° (Figure 10), being non-planar, decreasing the crystal packing energy and melting point, thus improving the aqueous solubility. In a search on CCDC version 5.41 (November 2019) the (E)-1-(4-bromophenyl)-3-phenylprop-2-en-1-one (BRCHAL) [49] was found, which has the chalcone moiety without the butyl group. It has two independent molecules in asymmetric unity with similar planarity (angles between their aromatic rings are 49.70° and 49.98° (Figure 11)); this structure was tested before with *Ae. aegypti*, and the results are presented in Table 3.

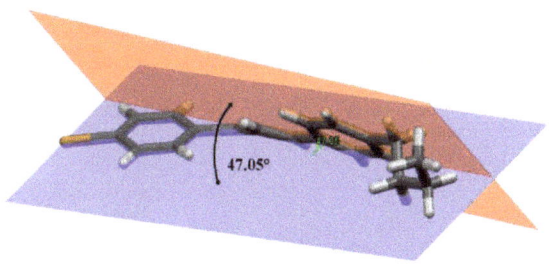

Figure 10. Illustration of the angles between rings for BBP.

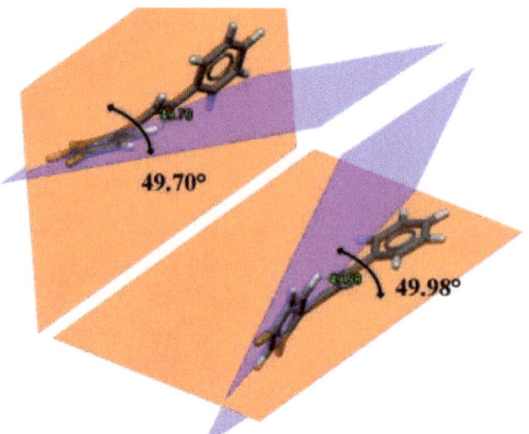

Figure 11. Illustration of the angles between rings for BRCHAL.

Table 3. Larvicidal activity of BBP on *A. aegypti* larvae (L3) after 48 h.

LC$_{25}$-48 h (mg·L^{-1}) *	LC$_{50}$-48 h (mg·L^{-1}) *	LC$_{75}$-48 h (mg·L^{-1}) *
37.7	46.0	54.2
(36.6–38.9)	(45.0–46.9)	(52.9–55.7)

* LC50, lethal concentration 50% mortality; LC25, lethal concentration 25% mortality; LC75, lethal concentration 75% mortality. Lethal concentrations in 48 h were calculated by non-linear regression of the respective dose–response curve, with a confidence interval of 95%. Values in parentheses represent lower and upper confidence limits of the interval.

The supramolecular arrangement of BPP is stabilized by C–H···O and the C–H···π interactions, with donor–acceptor distances being shorter than the sum of the van der Waals radii (2.7 Å for O and H) [50]. The BPP crystal packing molecules in a layer along the a-axis are presented in Figure 8 and were responsible for joining the crystal by the C5–H5···O1 interaction (d/Å, θ/°: 2.67 Å, 148°), symmetry code, (i) x − 1/2, −y + 1/2, −z + 1, and was also stabilized by edge-to-face C12–H12···π interaction involving aromatic ring B, shown in Figure 12 [29]. This was confirmed by the shape index (I) that provides information regarding π···π and C–H···π, and dnorm surface(II), with C–H···O observed by the concave red region over aromatic ring B. Such interactions also explain the observed conformation of the molecular structure at BBP [25,27].

HS indicates the partition of crystal electron density into molecular fragments from the three-dimensional space occupied by a molecule, representing the regions of space in which molecules are in contact with each other [33]. Both donor and acceptor regions are considered by the dnorm surface, which is based on the normalized function of di (distance to the nearest nucleus internal to the surface) and de (distance from the point to the nearest nucleus external to the surface) [51]. The combination of di and de on a two-dimensional plot (fingerprint) in Figure 13 indicates the nature of intermolecular interaction [51]. The major interaction lies in H–H (50.04%) and C–H (21.6%), while the halogen Br gives (12.4%) of interaction with H, and C–H (21.6%). The H–H contacts are the closest contacts for HS [52].

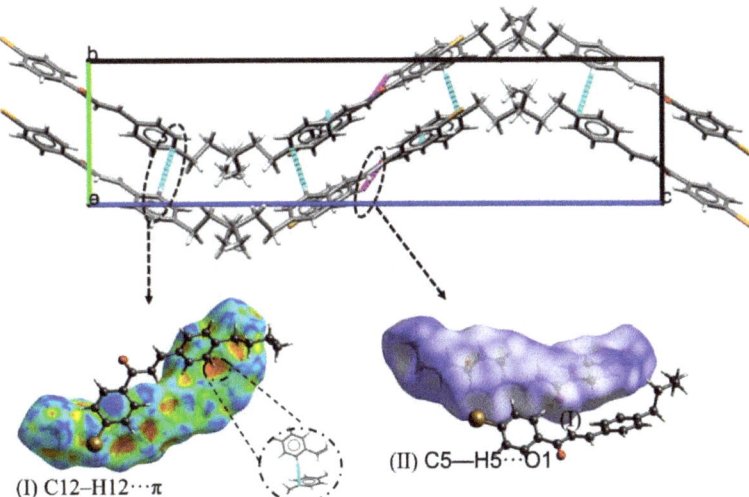

Figure 12. Present the packing stabilized by C–H···O and the C–H···π. Interactions that are confirmed by Hirshfeld surface shape index (**I**) showing the edge-to-face C–H···π interactions and the Hirshfeld surface dnorm map (**II**) showing C–H···O interactions for BBP.

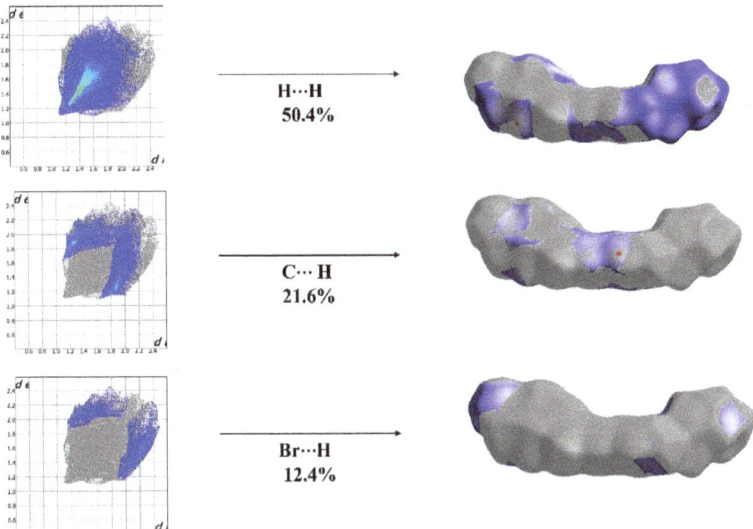

Figure 13. Hirshfeld surface diagram and two-dimensional fingerprint plots of the nearest internal distance (di) versus the nearest external distance (de) for major interactions in BBP.

The MEP map can be visualized reflecting the boundaries of the molecule, and it is useful to identify sites such as hydrogen bonds and regions where electrophilic and also nucleophilic attacks occur. This information can be used by electrostatic forces acting over long distances, exploring drug–receptor and enzyme–substrate interactions [53]. The molecular electrostatic potential V(r) at a point r is defined as Equation (1):

$$V(r) = \sum_i \frac{Z_i}{|R_i - r|} \int \left(\frac{\rho(r')}{|r' - r|}\right) dr' \qquad (2)$$

where Z_i is the charge at nucleus I located at R, and $\rho(r')$ is the molecular electron density at a point r' near the molecule [54]. Figure 14 shows the calculated MEP of BBP, showing regions that are well distinguished. Regions with negative electrostatic potential, i.e., C=O groups, are shown in red; regions with positive electrostatic potential are shown in blue, which means the region around the hydrogen atoms of the α,β-unsaturated ketone. This site tends to attract chemical species that are rich in electrons, while species poor in electrons will be attracted to the C=O region [37,40].

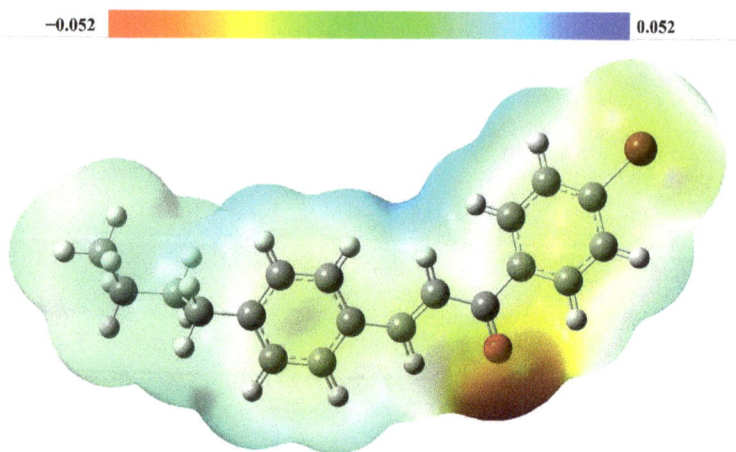

Figure 14. Molecular electrostatic potential map of BBP. The red region is rich in electrons, and the blue is electron depleted.

The frontier molecular orbitals (FMO) also play an important role in understanding chemical reaction mechanisms [42]. The LUMO and HOMO of other reactants interact with each other during the chemical process [53]. They are as follows: ELUMO = −2.57eV; EHOMO = −6.54eV. Their representations are presented in Figure 15.

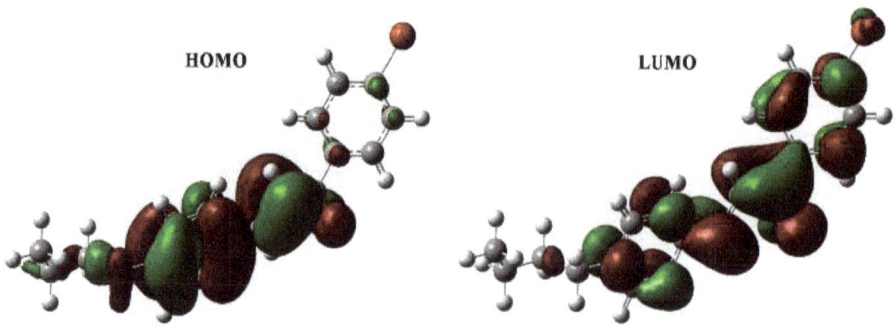

Figure 15. Graphical representation of the LUMO and HOMO and the calculated energy gap for BBP.

The energy gap between HOMO and LUMO is an important index for the stability of molecules, and a small gap automatically means small excitation energies for the manifold of excited states. Therefore, soft molecules, with a small gap, are more polarizable than hard molecules [55]. The energy values for these properties are: EGAP = 3.97 eV, hardness η = 1.78 eV, and softness σ = 0.56 eV. Comparing EGAP from BBP with other chalcone publications, ((E)-3-(4-ethoxyphenyl)-1-(4-methoxyphenyl)prop-2-en-1-one) has

EGAP = 6.4007 eV, and ((E)-1-(4-methoxyphenyl)-3-(3,4,5-trimethoxyphenyl) prop-2-en-1-one) EGAP = 6.2611 eV [56] as well as (E)-1-(2-aminophenyl)-3-(4-nitrophenyl)prop-2-en-1-one, with EGAP= 4.70 eV [57], but BBP presents a smaller EGAP value of 3.97 eV, so it is more reactive than these other chalcones.

3.3. Larvicidal Activity

According to The World Health Organization, larvicidal compounds should be active at concentrations value below 100 mg·L^{-1} [18,20,21]. Table 3 presents the larvicidal activity (LC25, LC50, and LC75) of BBP after 48 h. According to these results, BBP was effective against third-instar larvae (L3) of *A. aegypti*, showing LC50-48h = 46 mg·L^{-1}.

Recent studies have identified chalcone derivatives as anti-juvenile compounds (antagonists of the juvenile insect hormone) [19]. These natural product derivatives are important interferences in the physiological processes of insects, such as interference in ecdysis, metamorphosis, and reproduction, among others [19]. In addition, our results showed that, as well as lethality among larvae of *Ae. aegypti*, there was abnormal larval development (inhibition of ecdysis) when larvae were exposed to BBP concentrations below 20 mg.L^{-1}. Although there is no lethality at these concentrations, the larvae exposed to BBP for 48 h showed a gradual reduction in their mobility and lethargy compared to the control. Figure 16 shows larval mortality (%) as a function of BBP concentration.

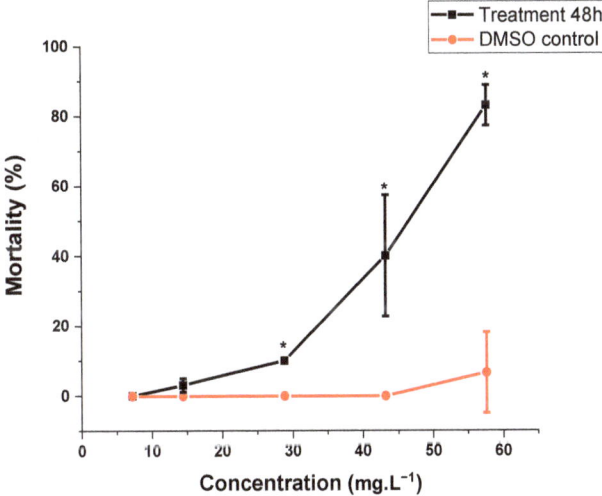

Figure 16. *Ae. aegypti* larvae mortality (%) as a function of BBP concentration (mg·L^{-1}). The symbol (*) indicates statistical significance with the DMSO group (ANOVA/Tukey; $p \leq 0.05$).

A mortality rate depends on the concentration of BBP used was observed, presenting a rate up to 80% at 57.6 mg L^{-1}, in 48 h (Figure 15). In comparison to the literature, the bromochalcone derivative (BBP) showed greater larvicidal activity than other bromochalcone derivatives evaluated by Kalirajan et al. (2015) on *Culex* sp. and *Anopheles* sp. [58]. Eight of these molecules showed LC50-24 h values greater than 50 mg·L^{-1}. Table 4 presents the structure–activity relationship of bromochalcone derivatives concerning *Ae. aegypti* larvae mortality (%).

Table 4. Structure–activity relationship of bromochalcone derivatives concerning A. aegypti larvae mortality (%).

Entry	R	Concentration (mg L^{-1})	Mortality (%)	Ref
1	H	100	70	[18]
2	NO$_2$	100	0	[18]
3	OMe	100	0	[18]
4	(CH$_2$)$_3$CH$_3$	54.2	75	This work

From the analysis of Table 4, there is a significant relationship between the molecular structure and the larvicidal activity of bromochalcone derivatives. Our results compared to the literature demonstrated that the presence of withdrawing groups e.g., (R = NO$_2$) and donating groups e.g., (R = OMe) decreased the larvicidal efficiency when compared to the unsubstituted chalcone (R = H). Additionally, that study revealed that the insertion of the butyl group at position 4 (R = (CH$_2$)$_3$CH$_3$) significantly increased the larvicidal activity of bromochalcone derivatives. This increase in the larvicidal activity of BBP may be related to the increase in its lipophilicity and interaction with biological membranes. The obtained results are promising, and bromochalcone derivatives may be a potential arsenal against Ae. aegypti larvae.

4. Conclusions

A bromochalcone derivative (BBP) was obtained in excellent yield and fully characterized by ^{1}H- and ^{13}C-NMR, IR, FT-Raman, MS, UV-Vis, TGA, and DSC. Moreover, according to the literature and our study, the substitution pattern of chalcone derivative influenced its larvicidal activity. In this regard, BBP (a bromo chalcone derivative), which presents a butyl group at position 4 (R = (CH$_2$)$_3$CH$_3$, significantly showed promising larvicidal activity (mortality up to 80% at 57.6 mg·L^{-1}) against Ae. aegypti larvae. BBP showed a non-planar molecular conformation with 2.9° of difference to BRCHAL(R=H) that presents 5% less mortality than BBP against Ae. aegypti larvae at 100 mg·L^{-1}, this difference between the planarity indicated a small improvement in the solubility in BBP compared with BRCHAL. The solid-state of BBP was stabilized by intermolecular interactions C–H···O and C–H···π, with theoretical calculations indicating electrophilic sites near oxygen atoms and depleted electrons around the hydrogen of the α,β-unsaturated ketone. The bandgap was 3.79, which is more polarizable than similar reported structures. Overall, we presented the synthesis, structural characterization, computational analysis, and full spectroscopy/thermal characterization of the BBP as well as its larvicidal activity against Ae. aegypti larvae.

Author Contributions: Conceptualization, P.P.F., L.D.D. and H.B.N.; methodology, J.E.Q., P.R.S.W., P.P.F. and W.F.V.; formal analysis P.P.F. and W.F.V.; investigation, P.P.F., W.F.V., L.M.d.S. and L.D.D.; resources, J.E.Q., I.I. and G.L.B.d.A.; data curation, P.P.F. and J.E.Q.; writing—original draft preparation, P.P.F., W.F.V., L.M.d.S., L.D.D. and J.M.F.C.; writing—review and editing, P.P.F., L.D.D. and P.R.S.W.; visualization, P.P.F. and L.D.D.; supervision H.B.N., G.L.B.d.A. and A.G.O.; funding acquisition, P.R.S.W. and H.B.N. All authors have read and agreed to the published version of the manuscript.

Funding: This research was funded by Fundação de Amparo à Pesquisa do Estado de São Paulo (FAPESP) CEPOF 2013/07276-1, Coordenação de Aperfeiçoamento de Pessoal de Nível Superior—Brasil (CAPES—Finance Code 001, Convênio n° 817164/2015 CAPES/PROAP), and INCT Basic Optics and Applied to Life Sciences (FAPESP 2014/50857-8, CNPq 465360/2014-9). P.P Firmino thanks Conselho Nacional de Desenvolvimento Científico e Tecnológico (CNPq 160856/2021-3). L. D. Dias thanks Fundação de Amparo à Pesquisa do Estado de São Paulo (FAPESP) for Post-doc grant 2019/13569-8. Research was developed with support from the High Performance Computing Center at Universidade Estadual de Goiás (UEG).

Institutional Review Board Statement: Not applicable.

Informed Consent Statement: Informed consent was obtained from all subjects involved in the study.

Data Availability Statement: Not applicable.

Conflicts of Interest: The authors declare no conflict of interest.

References

1. Cox, C.D.; Breslin, M.J.; Mariano, B.J. Two-Step Synthesis of β-Alkyl Chalcones and Their Use in the Synthesis of 3,5-Diaryl-5-Alkyl-4,5-Dihydropyrazoles. *Tetrahedron Lett.* **2004**, *45*, 1489–1493. [CrossRef]
2. Xu, M.; Wu, P.; Shen, F.; Ji, J.; Rakesh, K.P. Chalcone Derivatives and Their Antibacterial Activities: Current Development. *Bioorganic Chem.* **2019**, *91*, 103133. [CrossRef] [PubMed]
3. Ngaini, Z.; Haris Fadzillah, S.M.; Hussain, H. Synthesis and Antimicrobial Studies of Hydroxylated Chalcone Derivatives with Variable Chain Length. *Nat. Prod. Res.* **2012**, *26*, 892–902. [CrossRef] [PubMed]
4. Stepanić, V.; Matijašić, M.; Horvat, T.; Verbanac, D.; Chlupáćová, M.K.; Saso, L.; Žarković, N. Antioxidant Activities of Alkyl Substituted Pyrazine Derivatives of Chalcones—In Vitro and in Silico Study. *Antioxidants* **2019**, *8*, 90. [CrossRef] [PubMed]
5. Da Silva, P.T.; da Cunha Xavier, J.; Freitas, T.S.; Oliveira, M.M.; Coutinho, H.D.M.; Leal, A.L.A.B.; Barreto, H.M.; Bandeira, P.N.; Nogueira, C.E.S.; Sena, D.M.; et al. Synthesis, Spectroscopic Characterization and Antibacterial Evaluation by Chalcones Derived of Acetophenone Isolated from Croton Anisodontus Müll.Arg. *J. Mol. Struct.* **2021**, *1226*, 129403. [CrossRef]
6. Yadav, V.R.; Prasad, S.; Sung, B.; Aggarwal, B.B. The Role of Chalcones in Suppression of NF-KB-Mediated Inflammation and Cancer. *Int. Immunopharmacol.* **2011**, *11*, 295–309. [CrossRef]
7. Wang, G.; Liu, W.; Gong, Z.; Huang, Y.; Li, Y.; Peng, Z. Synthesis, Biological Evaluation, and Molecular Modelling of New Naphthalene-Chalcone Derivatives as Potential Anticancer Agents on MCF-7 Breast Cancer Cells by Targeting Tubulin Colchicine Binding Site. *J. Enzym. Inhib. Med. Chem.* **2020**, *35*, 139–144. [CrossRef]
8. Bonakdar, A.P.S.; Vafaei, F.; Farokhpour, M.; Esfahani, M.H.N.; Massah, A.R. Synthesis and Anticancer Activity Assay of Novel Chalcone-Sulfonamide Derivatives. *Iran. J. Pharm. Res.* **2017**, *16*, 565–568. [CrossRef]
9. Mellado, M.; Espinoza, L.; Madrid, A.; Mella, J.; Chávez-Weisser, E.; Diaz, K.; Cuellar, M. Design, Synthesis, Antifungal Activity, and Structure–Activity Relationship Studies of Chalcones and Hybrid Dihydrochromane–Chalcones. *Mol. Divers.* **2019**, *24*, 603–615. [CrossRef]
10. Zheng, Y.; Wang, X.; Gao, S.; Ma, M.; Ren, G.; Liu, H.; Chen, X. Synthesis and Antifungal Activity of Chalcone Derivatives. *Nat. Prod. Res.* **2015**, *29*, 1804–1810. [CrossRef]
11. Lahsasni, S.A.; Al Korbi, F.H.; Aljaber, N.A.A. Synthesis, Characterization and Evaluation of Antioxidant Activities of Some Novel Chalcones Analogues. *Chem. Cent. J.* **2014**, *8*, 32. [CrossRef]
12. Díaz-Carrillo, J.T.; Díaz-Camacho, S.P.; Delgado-Vargas, F.; Rivero, I.A.; López-Angulo, G.; Sarmiento-Sánchez, J.I.; Montes-Avila, J. Synthesis of Leading Chalcones with High Antiparasitic, against Hymenolepis Nana, and Antioxidant Activities. *Braz. J. Pharm. Sci.* **2018**, *54*, 1–13. [CrossRef]
13. AID 371504—Antiamnesic Activity against Entamoeba Histolytica HMI:IMSS after 72 Hrs by Microdilution Method—PubChem. Available online: https://pubchem.ncbi.nlm.nih.gov/bioassay/371504#sid=103590847 (accessed on 29 October 2020).
14. Chhabra, M.; Mittal, V.; Bhattacharya, D.; Rana, U.V.S.; Lal, S. Chikungunya Fever: A Re-Emerging Viral Infection. *Indian J. Med. Microbiol.* **2008**, *26*, 5–12. [CrossRef]
15. WHO. Dengue and Severe Dengue. Available online: https://www.who.int/news-room/fact-sheets/detail/dengue-and-severe-dengue (accessed on 18 February 2021).
16. Ghosh, A.; Dar, L. Dengue Vaccines: Challenges, Development, Current Status and Prospects. *Indian J. Med. Microbiol.* **2015**, *33*, 3–15. [CrossRef] [PubMed]
17. Barbosa, P.B.B.M.; de Oliveira, J.M.; Chagas, J.M.; Rabelo, L.M.A.; de Medeiros, G.F.; Giodani, R.B.; da Silva, E.A.; Uchôa, A.F.; de Fátima de Freire Melo Ximenes, M. Evaluation of Seed Extracts from Plants Found in the Caatinga Biome for the Control of Aedes Aegypti. *Parasitol. Res.* **2014**, *113*, 3565–3580. [CrossRef] [PubMed]
18. Targanski, S.K.; Sousa, J.R.; de Pádua, G.M.S.; de Sousa, J.M.; Vieira, L.C.C.; Soares, M.A. Larvicidal Activity of Substituted Chalcones against Aedes Aegypti (Diptera: Culicidae) and Non-Target Organisms. *Pest Manag. Sci.* **2021**, *77*, 325–334. [CrossRef]

19. Lee, S.H.; Oh, H.W.; Fang, Y.; An, S.B.; Park, D.S.; Song, H.H.; Oh, S.R.; Kim, S.Y.; Kim, S.; Kim, N.; et al. Identification of Plant Compounds That Disrupt the Insect Juvenile Hormone Receptor Complex. *Proc. Natl. Acad. Sci. USA* **2015**, *112*, 1733–1738. [CrossRef]
20. Cheng, S.S.; Chang, H.T.; Chang, S.T.; Tsai, K.H.; Chen, W.J. Bioactivity of Selected Plant Essential Oils against the Yellow Fever Mosquito Aedes Aegypti Larvae. *Bioresour. Technol.* **2003**, *89*, 99–102. [CrossRef]
21. WHO and Special Programme for research and Training in Tropical Diseases. *Dengue Guidelines for Diagnosis, Treatment, Prevention and Control*; World Health Organization: Geneva, Switzerland, 2009; pp. 1–144. ISBN 9789241547871.
22. Ternavisk, R.R.; Camargo, A.J.; Machado, F.B.C.; Rocco, J.A.F.F.; Aquino, G.L.B.; Silva, V.H.C.; Napolitano, H.B. Synthesis, Characterization, and Computational Study of a New Dimethoxy-Chalcone. *J. Mol. Modeling* **2014**, *20*, 2526. [CrossRef]
23. Pence, I.; Mahadevan-jansen, A. Clinical instrumentation and applications of Raman spectroscopy. *Chem. Soc. Rev.* **2016**, *45*, 1958–1979. [CrossRef]
24. Mitsutake, H.; Poppi, R.J.; Breitkreitz, M.C. Raman Raman Imaging Imaging Spectroscopy: Spectroscopy: History, History, Fundamentals Fundamentals and Current and Current Scenario Scenario of the Oftechnique the Technique. *J. Braz. Chem. Soc.* **2019**, *30*, 2243–2258. [CrossRef]
25. Bolton, E.E.; Wang, Y.; Thiessen, P.A.; Bryant, S.H. *Chapter 12 PubChem: Integrated Platform of Small Molecules and Biological Activities*; Elsevier B.V.: Amsterdam, The Netherlands, 2008; Volume 4, ISBN 9780444532503.
26. Sheldrick, G.M. A Short History of SHELX. *Acta Crystallogr. Sect. A Found. Crystallogr.* **2008**, *64*, 112–122. [CrossRef]
27. Dolomanov, O.V.; Bourhis, L.J.; Gildea, R.J.; Howard, J.A.K.; Puschmann, H. OLEX2: A Complete Structure Solution, Refinement and Analysis Program. *J. Appl. Crystallogr.* **2009**, *42*, 339–341. [CrossRef]
28. Sheldrick, G.M. Crystal Structure Refinement with SHELXL. *Acta Crystallogr. Sect. C Struct. Chem.* **2015**, *71*, 3–8. [CrossRef] [PubMed]
29. Spek, A.L. Structure Validation in Chemical Crystallography. *Acta Crystallogr. Sect. D Biol. Crystallogr.* **2009**, *65*, 148–155. [CrossRef]
30. Kanagasabai, S.; Gugan, K.; Reddy, M.B.; Anandhan, R.; Kumar, S.M.; Usharani, S. Crystal Structure and Hirshfeld Surface Analysis of 2,2'-Bi-(3-Phenyl-2H-1,4-Benzothiazine). *Chem. Data Collect.* **2019**, *20*, 1564–1567. [CrossRef]
31. Abad, N.; Ramli, Y.; Hökelek, T.; Sebbar, N.K.; Mague, J.T.; Essassi, E.M. Crystal Structure and Hirshfeld Surface Analysis of Ethyl 2-{4-[(3-Methyl-2-Oxo-1,2-Dihydroquinoxalin- 1-Yl)Methyl]-1H-1,2,3-Triazol-1-Yl}acetate. *Acta Crystallogr. Sect. E Crystallogr. Commun.* **2018**, *74*, 1648–1652. [CrossRef]
32. McKinnon, J.J.; Jayatilaka, D.; Spackman, M.A. Towards Quantitative Analysis of Intermolecular Interactions with Hirshfeld Surfaces. *Chem. Commun.* **2007**, 3814–3816. [CrossRef]
33. Spackman, M.A.; Jayatilaka, D. Hirshfeld Surface Analysis. *CrystEngComm* **2009**, *11*, 19–32. [CrossRef]
34. Spackman, P.R.; Turner, M.J.; McKinnon, J.J.; Wolff, S.K.; Grimwood, D.J.; Jayatilaka, D.; Spackman, M.A. CrystalExplorer: A Program for Hirshfeld Surface Analysis, Visualization and Quantitative Analysis of Molecular Crystals. *J. Appl. Crystallogr.* **2021**, *54*, 1006–1011. [CrossRef]
35. Frisch, M.J.; Trucks, G.W.; Schlegel, H.B.; Scuseria, G.E.; Robb, M.A.; Cheeseman, J.R.; Scalmani, G.; Barone, V.; Mennucci, B.; Petersson, G.A.; et al. Gaussian 09 2009. Available online: https://gaussian.com/gaussian16/ (accessed on 28 February 2022).
36. Custodio, J.M.F.; Michelini, L.J.; De Castro, M.R.C.; Vaz, W.F.; Neves, B.J.; Cravo, P.V.L.; Barreto, F.S.; Filho, M.O.M.; Perez, C.N.; Napolitano, H.B. Structural Insights into a Novel Anticancer Sulfonamide Chalcone. *New J. Chem.* **2018**, *42*, 3426–3434. [CrossRef]
37. Sallum, L.O.; Siqueira, V.L.; Custodio, J.M.F.; Borges, N.M.; Lima, A.P.; Abreu, D.C.; De Lacerda, E.P.S.; Lima, R.S.; De Oliveira, A.M.; Camargo, A.J.; et al. Molecular Modeling of Cytotoxic Activity of a New Terpenoid-like Bischalcone. *New J. Chem.* **2019**, *43*, 18451–18460. [CrossRef]
38. Haroon, M.; Khalid, M.; Akhtar, T.; Tahir, M.N.; Khan, M.U.; Saleem, M.; Jawaria, R. Synthesis, Spectroscopic, SC-XRD Characterizations and DFT Based Studies of Ethyl2-(Substituted-(2-Benzylidenehydrazinyl))Thiazole-4-Carboxylate Derivatives. *J. Mol. Struct.* **2019**, *1187*, 164–171. [CrossRef]
39. Moreira, C.A.; Custódio, J.M.F.; Vaz, W.F.; D'Oliveira, G.D.C.; Noda Perez, C.; Napolitano, H.B. A Comprehensive Study on Crystal Structure of a Novel Sulfonamide-Dihydroquinolinone through Experimental and Theoretical Approaches. *J. Mol. Modeling* **2019**, *25*, 205. [CrossRef]
40. Medina, D.; Menezes, A.C.S.; Sousa, J.E.F.; Oliveira, S.S. Structural and Theoretical Investigation of Anhydrous 3, 4, 5-Triacetoxybenzoic Acid. *PLoS ONE* **2016**, *11*, e0158029. [CrossRef]
41. Merrick, J.P.; Moran, D.; Radom, L. An Evaluation of Harmonic Vibrational Frequency Scale Factors. *J. Phys. Chem. A* **2007**, *111*, 11683–11700. [CrossRef]
42. Aguiar, A.S.N.; Queiroz, J.E.; Firmino, P.P.; Vaz, W.F.; Camargo, A.J.; de Aquino, G.L.B.; Napolitano, H.B.; Oliveira, S.S. Synthesis, Characterization, and Computational Study of a New Heteroaryl Chalcone. *J. Mol. Modeling* **2020**, *26*, 243. [CrossRef]
43. Millam, R.D.; Todd, A.; Keith John, M. GaussView Version 6. 2019. Available online: https://gaussian.com/gaussview6/ (accessed on 28 February 2022).
44. Cottrell, S.J.; Olsson, T.S.G.; Taylor, R.; Cole, J.C.; Liebeschuetz, J.W. Validating and Understanding Ring Conformations Using Small Molecule Crystallographic Data. *J. Chem. Inf. Modeling* **2012**, *52*, 956–962. [CrossRef]

45. Bourhis, L.J.; Dolomanov, O.V.; Gildea, R.J.; Howard, J.A.K.; Puschmann, H. The Anatomy of a Comprehensive Constrained, Restrained Refinement Program for the Modern Computing Environment—Olex2 Dissected. *Acta Crystallogr. Sect. A Found. Crystallogr.* **2015**, *71*, 59–75. [CrossRef]
46. Sheldrick, G.M. SHELXT—Integrated Space-Group and Crystal-Structure Determination. *Acta Crystallogr. Sect. A Found. Crystallogr.* **2015**, *71*, 3–8. [CrossRef]
47. Custodio, J.M.F.; Gotardo, F.; Vaz, W.F.; D'Oliveira, G.D.C.; de Almeida, L.R.; Fonseca, R.D.; Cocca, L.H.Z.; Perez, C.N.; Oliver, A.G.; de Boni, L.; et al. Benzenesulfonyl Incorporated Chalcones: Synthesis, Structural and Optical Properties. *J. Mol. Struct.* **2020**, *1208*, 127845. [CrossRef]
48. Rajesh Kumar, P.C.; Ravindrachary, V.; Janardhana, K.; Manjunath, H.R.; Karegouda, P.; Crasta, V.; Sridhar, M.A. Optical and Structural Properties of Chalcone NLO Single Crystals. *J. Mol. Struct.* **2011**, *1005*, 1–7. [CrossRef]
49. Rabinovich, D.; Schmidt, G.M.J.; Shaked, Z. Topochemistry. Part XXXIV. Crystal and Molecular Structure of P′-Bromochalcone. *J. Chem. Soc. Perkin Trans. 2* **1973**, *1*, 33–37. [CrossRef]
50. Johnston, R.C.; Cheong, P.H.Y. C-H···O Non-Classical Hydrogen Bonding in the Stereomechanics of Organic Transformations: Theory and Recognition. *Org. Biomol. Chem.* **2013**, *11*, 5057–5064. [CrossRef] [PubMed]
51. Vaz, W.F.; Custodio, J.M.F.; D'Oliveira, G.D.C.; Neves, B.J.; Junior, P.S.C.; Filho, J.T.M.; Andrade, C.H.; Perez, C.N.; Silveira-Lacerda, E.P.; Napolitano, H.B. Dihydroquinoline Derivative as a Potential Anticancer Agent: Synthesis, Crystal Structure, and Molecular Modeling Studies. *Mol. Divers.* **2021**, *25*, 55–66. [CrossRef]
52. Mkaouar, I.; Karâa, N.; Hamdi, B.; Zouari, R. Synthesis, Crystal Structure, Thermal Analysis, Vibrational Study Dielectric Behaviour and Hirshfeld Surface Analysis of $[C_6H_{10}(NH_3)_2)]_2$ $SnCl_6$ $(Cl)_2$. *J. Mol. Struct.* **2016**, *1115*, 161–170. [CrossRef]
53. Oliveira, S.S.; Santin, L.G.; Almeida, L.R.; Malaspina, L.A.; Lariucci, C.; Silva, J.F.; Fernandes, W.B.; Aquino, G.L.B.; Gargano, R.; Camargo, A.J.; et al. Synthesis, Characterization, and Computational Study of the Supramolecular Arrangement of a Novel Cinnamic Acid Derivative. *J. Mol. Modeling* **2017**, *23*, 35. [CrossRef]
54. Chidangil, S.; Shukla, M.K.; Mishra, P.C. A Molecular Electrostatic Potential Mapping Study of Some Fluoroquinolone Anti-Bacterial Agents. *J. Mol. Modeling* **1998**, *4*, 250–258. [CrossRef]
55. Vaz, W.F.; Custodio, J.M.F.; Silveira, R.G.; Castro, A.N.; Campos, C.E.M.; Anjos, M.M.; Oliveira, G.R.; Valverde, C.; Baseia, B.; Napolitano, H.B. Synthesis, Characterization, and Third-Order Nonlinear Optical Properties of a New Neolignane Analogue. *RSC Adv.* **2016**, *6*, 79215–79227. [CrossRef]
56. Custodio, J.; Faria, E.; Sallum, L.; Duarte, V.; Vaz, W.; de Aquino, G.; Carvalho Jr., P.; Napolitano, H. The Influence of Methoxy and Ethoxy Groups on Supramolecular Arrangement of Two Methoxy-Chalcones. *J. Braz. Chem. Soc.* **2017**, *28*, 2180–2191. [CrossRef]
57. Michelini, L.J.; Castro, M.R.C.; Custodio, J.M.F.; Naves, L.F.N.; Vaz, W.F.; Lobón, G.S.; Martins, F.T.; Perez, C.N.; Napolitano, H.B. A Novel Potential Anticancer Chalcone: Synthesis, Crystal Structure and Cytotoxic Assay. *J. Mol. Struct.* **2018**, *1168*, 309–315. [CrossRef]
58. Kalirajan, R.; Jubie, S.; Gowramma, B. Microwave Irradated Synthesis, Characterization and Evaluation for Their Antibacterial and Larvicidal Activities of Some Novel Chalcone and Isoxazole Substituted 9-Anilino Acridines. *Open J. Chem.* **2015**, *1*, 1–7. [CrossRef]

Article

Isolation, Identification, Spectral Studies and X-ray Crystal Structures of Two Compounds from *Bixa orellana*, DFT Calculations and DNA Binding Studies

Mehtab Parveen [1,*], Mohammad Azeem [1], Afroz Aslam [1], Mohammad Azam [2,*], Sharmin Siddiqui [3], Mohammad Tabish [3], Ali Mohammad Malla [1,4], Kim Min [5], Vitor Hugo Rodrigues [6], Saud I. Al-Resayes [2] and Mahboob Alam [5,*]

1. Division of Organic Synthesis, Department of Chemistry, Aligarh Muslim University, Aligarh 202002, India; azam23961@gmail.com (M.A.); afrozaslam10@gmail.com (A.A.); haiderchem09@gmail.com (A.M.M.)
2. Department of Chemistry, College of Science, King Saud University, P.O. Box 2455, Riyadh 11451, Saudi Arabia; sresayes@ksu.edu.sa
3. Department of Biochemistry, Aligarh Muslim University, Aligarh 202002, India; sharminbcb@gmail.com (S.S.); tabish.biochem@gmail.com (M.T.)
4. Government Degree College, Beerwah, Budgam, Kashmir 193411, India
5. Department of Safety Engineering, Dongguk University, 123 Dongdae-ro, Gyeongju 780714, Gyeongbuk, Korea; kimmin@dongguk.ac.kr
6. CFisUC, Department of Physics, University of Coimbra, 3004-516 Coimbra, Portugal; vhugo@uc.pt
* Correspondence: mehtab.organic2009@gmail.com (M.P.); mhashim@ksu.edu.sa (M.A.); mahboobchem@gmail.com (M.A.)

Citation: Parveen, M.; Azeem, M.; Aslam, A.; Azam, M.; Siddiqui, S.; Tabish, M.; Malla, A.M.; Min, K.; Rodrigues, V.H.; Al-Resayes, S.I.; et al. Isolation, Identification, Spectral Studies and X-ray Crystal Structures of Two Compounds from *Bixa orellana*, DFT Calculations and DNA Binding Studies. *Crystals* 2022, 12, 380. https://doi.org/10.3390/cryst12030380

Academic Editors: Radu Claudiu Fierascu, Florentina Monica Raduly and Dinadayalane Tandabany

Received: 10 February 2022
Accepted: 7 March 2022
Published: 11 March 2022

Publisher's Note: MDPI stays neutral with regard to jurisdictional claims in published maps and institutional affiliations.

Copyright: © 2022 by the authors. Licensee MDPI, Basel, Switzerland. This article is an open access article distributed under the terms and conditions of the Creative Commons Attribution (CC BY) license (https://creativecommons.org/licenses/by/4.0/).

Abstract: 4,6-Diacetylresorcinol (**1**) and 3-*O*-methylellagic acid dihydrate (**2**), both biologically significant compounds, were extracted from *Bixa orellana* and studied using IR, ^{1}H, and ^{13}C NMR, and UV-vis spectroscopic techniques. X-ray crystallographic techniques were also used to establish the molecular structure of the isolated compounds **1** and **2**. Geometric parameters, vibrational frequencies, and gauge including atomic orbital (GIAO) ^{1}H and ^{13}C NMR of **1** and **2** in the ground state were computed by the density functional theory (DFT) using B3LYP/6-311G(d,p) basis set backing up experimental studies and established the correct structure of isolated compounds. The parameters obtained from the combined DFT, and X-ray diffraction studies are mutually agreed to establish correct structures of **1** and **2**. In addition, an electrostatic potential map and HOMO−LUMO energy gap were made using the DFT calculation to determine the distribution of energy and the chemical reactivity region of the isolated compounds. The current study also provides further insights into the interaction of compound **2** with ct-DNA using numerous biophysical and *in silico* techniques. Moreover, *in silico* studies indicate that compound **2** binds to the DNA in the minor groove. Lipinski's rule of five revealed a higher tendency of compound **2** towards drug-likeness. The bioavailability and synthetic accessibility score for compound **2** was found to be 0.55 and 3.21, suggesting that compound **2** could serve as an effective therapeutic candidate.

Keywords: *Bixa orellana* (Family: Bixaceae); 4,6-diacetylresorcinol; 3-*O*-methylellagic acid dihydrate; X-ray diffraction; DFT; NMR; frontier molecular orbitals

1. Introduction

The Bixaceae family includes *Bixa orellana* L., also known as annatto. It is a Central and South American shrub that grows 3–6 m tall and is one of the oldest plants to produce natural colors. Originally, the herb was used as body paint, a treatment for heartburn, an insect repellant, a sunscreen, and to fight off evil. It was named after *Francisco de Orellana*, a Spanish conquistador [1]. Constipation, fevers, heartburn, asthma, scabies, ulcers, diarrhea, stomach upset, skin diseases, measles, anecdotal treatment of diabetes, allergy, leprosy, infectious diseases, burns, measles, gonorrhea, diarrhea, asthma, angina, tumors, skin problems, and urinary infections have all been treated with *Bixa orellana* in different areas of the globe (oral and topic) [2,3] for centuries. Indigenous people have conventionally utilized the pulp from this plant's seeds externally to increase lips radiance, which is how *Bixa orellana* obtained its moniker, "lipstick tree" [4]. *Bixa orellana* is widely used in dairy food coloring and bleaching, specifically bakery products, cream desserts, buttermilk deserts, rice flour, and corn starch [5–7]. In recent decades, several distinct groups of phytoconstituents, including aliphatic compounds, carotenoids, apocarotenoids, sterols, volatile oils, monoterpenes, sesquiterpenes, triterpenoids, and other miscellaneous substances, have been isolated from all parts of this plant [8–10].

In continuance of our ongoing research [11], we herein report the extraction, isolation, spectroscopic, X-ray crystallographic techniques of two compounds [4-acetylresorcinol (**1**), 3-O-methylellagic acid dihydrate (**2**)], derived from the *Bixa Orellana* leaves. Physiochemical spectral data (FTIR, UV, ^{1}H NMR, ^{13}C NMR, and MS spectral analysis), including Single Crystal X-ray Diffraction was used to establish the chemical structure of two isolated phytoconstituents. Compound **1** has not yet been isolated out of this plant source, to our knowledge. The structure of **2**, which crystallized as white crystals, was validated by X-ray crystallography. Quantum chemical calculations using the DFT (B3LYP) theory have yet to be performed on these molecules **1** and **2**. The B3LYP functional was adopted because to its significance in quantum chemistry as well as its precision Geometry calculation, IR, NMR spectra, and a variety of other molecular properties for compounds were investigated in this study at the B3LYP/6-311(d,p) level of theory. The FTIR vibrational bands are interpreted using harmonic force field calculation without scaling, which has a lower computational cost and is sometimes important in extremely large chemical substances. Gauge independent atomic orbital (GIAO) approach is used to analyze NMR spectra (^{1}H and ^{13}C) in gaseous and solvent phases at the same level of theory. Theoretical results were compared to experimental data and found to be in good agreement.

2. Materials and Methods

2.1. Reagents and Apparatus

Silica gel (60–120 mesh) used for various chromatographic procedures was supplied by Merck (India) and Merck (Germany). The various solvents systems employed for thin layer chromatography were benzene-chloroform (8:2) and petroleum ether (60–80 °C)-benzene (1:1). A Kofler block was used to record melting points of compounds and were uncorrected. Carlo Erba analyzer model 1108 was used for elemental analysis (C, H, N). The IR spectra on KBr pellets were obtained using Interspec-2020 (FTIR) and Shimadzu IR-408 Perkin-Elmer 1800 (FTIR). UV-Vis recordings in methanolic solution were made with a Shimadzu UV-1800 UV-Vis spectrophotometer (Shimadzu, Japan). ^{1}H NMR and ^{13}C NMR spectra were performed in $CDCl_3$/DMSO-d_6 with TMS as an internal standard on Bruker Avance II 400 MHz spectrometers. Chemical shifts were expressed in ppm (δ). In the MS (EI) mode, mass spectra were recorded on JEOL D−300 mass spectrometer. Iodine vapor was used to verify purity for the compounds by observing spots on TLC.

2.2. Plant Material

Dr. Athar Ali Khan, Taxonomist, Department of Botany, Aligarh Muslim University, Aligarh, identified a specimen plant of *Bixa orellana* L. (Family: Bixaceae) taken from Allahabad University in Allahabad, India. A specimen with voucher number 243 was registered at the Botany Department, AMU, Aligarh.

2.3. Extraction and Isolation

The leaves of *Bixa orellana* (Family: Bixaceae) were shade-dried and ground into powder (2 kg). A dark green gummy mass was obtained after extracting the air-dried powdered leaves 3 times with 95% ethanol at reflux temperature and evaporating the solvent under reduced pressure. Petroleum ether (60–80), benzene, ethyl acetate, acetone, and methanol were applied to extract the dark green gummy mass. Distillation was used to extract the solvent. Under reduced pressure, the petroleum ether and benzene extracts were concentrated, yielding a greenish sticky substance. The TLC study revealed that petroleum ether and benzene extracts acted the same way in different solvents, thus extracts were mixed. To elute the mixed petroleum ether–benzene extract from a silica gel column, a gradient of increasing solvents was used, including petroleum ether, petroleum ether–benzene, benzene, ethyl acetate, and eventually methanol. Those fractions with similar TLC characteristics and similar IR spectra were grouped. To obtain pure compounds 1 and 2, they were purified using repeated column chromatography followed by fractional crystallization and obtained 4,6-diacetylresorcinol (compound 1) and 3-O-methylalgeic acid dihydrate (Compound 2) was identified by comparison of m.p., TLC, Co-TLC and spectral data (IR, ^{1}H NMR, ^{13}C NMR and Mass) of authentic samples.

2.4. Spectral Analysis of Isolated Compounds

(1) 4,6-diacetylresorcinol: Colorless shining crystals (40 mg), m.p. 164–166 °C (lit. [12] 164–166 °C). Anal. Calc. for $C_{10}H_{10}O_4$; C, 61.85; H, 5.19; found: C, 61.83; H, 5.21. UV (MeOH) λ_{max}: 212, 275 and 312 nm. IR (KBr, ν_{max} cm^{-1}): 1589, 1490 (C=C), 1658 (C=O), 3430 (OH). ^{1}H NMR (400 MHz, DMSO-d_6, ppm) δ: 2.79 (s, 6H, 2 × COCH$_3$), 6.43 (s, 1H, H-2), 8.21 (s, 1H, H-5), 12.91 (s, 2H, 2 × OH); ^{13}C NMR (100 MHz, DMSO-d_6, ppm) δ: 26.12 (2 × CH$_3$), 104.99 (C-5), 113.66 (C-3), 136.31 (C-2), 168.93 (C-4, C-6), 202.52 (C-1); MS (ESI) (m/z): 194.06 [M$^+$] ($C_{10}H_{10}O_4$).

(2) 3-O-methylellagic acid dihydrate: Ethyl acetate fraction yielded compound 2. Crystallization from chloroform and methanol gave greenish white crystals (60 mg), m.p. 350–352 °C (lit. [11]). Anal. Calc. for $C_{15}H_{12}O_{10}$; C, 51.14; H, 3.41; found: C, 51.13; H, 3.43. UV (MeOH) λ_{max}: 245–400 nm. IR (KBr) ν cm^{-1}: 3310 (OH), 1721 (C=O), 1450 (C=C), 1171 (C-O). ^{1}H NMR (400 MHz, DMSO-d_6, ppm) δ: 10.59 (s, 2H, C-4 OH, C-4' OH), 8.72 (s, 1H, C-3' OH), 7.51 (s, 1H, H-5), 7.46 (s, 1H, H-5'), 4.15 (s, 3H, OCH$_3$). ^{13}C NMR (100 MHz, DMSO-d_6) δ: 159.02–159.03 (C-7 and C-7'), 154.10–154.14 (C-4 and C-4'), 143.24 (C-3'), 142.02–142.04 (C-2 and C-2'), 141.35 (C-3), 113.20–113.22 (C-6 and C-6'), 112.02–112.04 (C-1 and C-1'), 111.10–111.12 (C-5 and C-5'), 62.56 (3'-OCH$_3$). MS (EI): (m/z) 352.41 [M$^{+\bullet}$] ($C_{15}H_{12}O_{10}$).

2.5. Crystallographic Analysis

A Bruker Kappa APEXII CCD X-ray diffractometer was used for X-ray diffraction of single crystals of compounds 1 and 2 at ambient temperature using graphite-based Mo-K radiation of monochromatic wavelength (λ = 0.71073 Å). SADABS [13] was used to make absorption adjustments. The direct technique of SHELXL−97 was used to identify the expected structure of compounds 1 and 2 and anisotropically refined (non-H atoms) using SHELXL−97's complete matrix least-squares on F2 [14]. The hydrogen atoms in the structures were placed in a superlative location and refined using isotropic displacement parameters. PLATON [15,16] and ORTEP−3 [17] were used to analyze the layout of the figures of the structures. The experimental parameters related to the monocrystalline X-ray examination of 1 and 2 are described in Table S1 (Supplementary Material).

2.6. Computational Procedure

The isolated compounds (1 and 2) from *Bixa orellana* in their ground electronic states were subjected to density functional theory (DFT) calculations using density functional B3LYP-GD3 (Becke's 3 parameter hybrid functional for the exchange part, the Lee–Yang–Parr (LYP) correlation function, and Empirical D3 Dispersion [18]) and 6-311G(d,p) basis set by Gaussian package [19] using WebMO interface [20]. Under a strict convergence criterion, the isolated compound geometries in Cartesian representation were completely optimized in the ground electronic state. The ground state optimized geometry of the molecule was used to detect harmonic vibrational frequencies at the same level of theory. As a result, no imaginary frequencies were found, leading to the discovery of a true minima on the potential energy surface. The visualization programs GaussView [21], ChemCraft [22], and GaussSum [23] were used to construct realistic images, electronic transitions and vibrational assignments using animated modes. In the solution phase (CHCl$_3$ for compound **1** and DMSO for compound **2**), the GIAO-B3LYP/6-311G(d,p) level of theory was employed to measure NMR spectra of compounds applying gauge independent atomic orbital (GIAO) approach. In addition, UV-Vis spectra of the isolated compounds **1** and **2** were simulated using polarizable continuum model (PCM). Excitation energy, oscillatory power, wavelength and HOMO–LUMO energy differences were obtained at the basis set of TD-B3LYP-D3/6-311G (d,p). Molecular electrostatic potential (MEP) diagrams of both isolated compounds were generated using same level theory applied for optimization of the isolated compounds in order to predict interactions with neighboring molecules in term of electrophilic and nucleophilic sites.

2.7. ct-DNA Binding Studies

2.7.1. Sample Preparation for DNA Binding Experiments

To make a 4 mM stock solution, compound **2** was dissolved in ethanol. The compound working solutions were formed in accordance with the requirements. To dissolve calf thymus DNA, 10 mM Tris HCl (pH 7.4) was added to the buffer (ct-DNA). The DNA was kept at 4 °C. To achieve a homogeneous solution, the solution was occasionally stirred. To evaluate dissolved calf thymus DNA purity, the (A_{260}/A_{280}) absorbance ratio was utilized. The absorbance ratio of solution was 1.81, indicating that it is clean and does not require to be purified further. DNA content was calculated using the extinction value of molar 6600 M^{-1}cm^{-1} at 260 nm for a single nucleotide.

2.7.2. UV-Visible Absorption Spectroscopy

UV-visible absorption spectrum of compound **2** was measured using a Shimadzu UV-1800 spectrophotometer. At a fixed concentration of 200 M compound **2**, the reaction mixture was titrated with sequential concentrations of ct-DNA (0–50 M). A 10 mM Tris HCl buffer was used to adjust the baseline (pH 7.4).

2.7.3. Steady State Fluorescence

An RF-6000 Shimadzu spectrofluorometer was used to obtain steady-state fluorescence spectra of compound **2**. Spectra were obtained by adding ct-DNA (0–125 M) to a constant concentration of compound **2** (200 M). Compound **2** has a 273 nm excitation wavelength. The emission wavelength was adjusted accordingly, but the widths of the excitation and emission slits remained at 5 nm. Stern–Volmer constant (K_{sv}), binding constant (K_b) and constant (K_q) of bimolecular enhancement were all calculated using the data.

2.7.4. Circular Dichroism (CD)

A JASCO J-1500 circular dichroism spectrophotometer was used to collect and analyze the structural changes in ct-DNA produced by compound **2** interaction. The samples, which contained 200 µM of ct-DNA alone and 200 µM of compound **2** complexed with ct-DNA (1:1), were photographed at 200 nm/min in 190–340 nm wavelength range. In each batch of the experiment, a 10 mM Tris HCl (pH 7.4) buffer was utilized, as well as any necessary background adjustments. An average of 2 scans was used to produce the results.

2.7.5. Docking Studies

Docking analysis of 3-O-methylellagic (**2**) acid dihydrate was performed using Autodock 4.0 to describe the binding mode of compound **2** to DNA. The crystallographic information file (cif) of compound **2** was converted to PDB format using Mercury (Cambridge Crystallographic Data Centre). The RCSB database was used to retrieve the crystal structure of the synthetic DNA dodecamer d(CpGpCpGpApApTpTpCpGpCpG) with Pdb ID:1BNA. Avogadro was used to minimize the energy of the structures. Before docking, water molecules were stripped from the structures, Kollman charges were introduced, and polar hydrogen was supplied. Docking consisted of 10 runs using the Lamarckian genetic approach. The default settings for all other parameters were used. The x, y, and z-axis dimensions of the grid box for compound **2** were set to 26, 30, and 48 Å, respectively. The docking was conducted with a grid spacing of 0.375. UCSF Chimera 1.01, PyMOL and Accelrys Discovery Studio 4.5 were used to investigate the docked pose in the lowest configuration in order to determine the likely binding mode of compound 2 with DNA.

2.7.6. Prediction of Drug-Likeness (Lipinski's Rule of Five)

A rule of thumb (Lipinski's rule) can be used to analyze the drug-likeness qualities of chemical compounds and aid in the differentiation of drug-like and non-drug molecules. The drug-likeness of compound **2** was determined by calculating physiochemical characteristics such as the octanol-water partition coefficient (log P), molecular weight (MW), molar refractivity, polar surface area, hydrogen bond donors and acceptors. The following requirements apply: (i) less than 500 Da molecular mass, (ii) less than 5 hydrogen bond donors, (iii) strong lipophilicity, (iv) fewer than 10 acceptors for hydrogen bonds, and (v) a molar refractivity of 40 to 130. A tested compound that meets 2 or more of these rules is referred to as a drug-like molecule. Using online software tools [24], Lipinski's rule of five was employed to profile molecules at pH 7.

2.7.7. Synthetic Accessibility and Bioavailability Score Prediction

The SwissADME web tool was used to predict the bioavailability and synthetic accessibility score of the isolated molecule **2**. A compound can only be an effective drug [25] if it has a high bioavailability score and a low synthetic accessibility score (1 = easy to synthesize; 10 = difficult to synthesize).

3. Results and Discussion

*3.1. Crystal Structure, Molecular Geometry and IR Spectral Analysis of **1** and **2***

The isolated compounds (Figure 1) have four molecules per unit cell (Z = 4) in a P 21/c crystal structure. From single crystal X-ray diffraction data, it is found that the crystal of both compounds (Figure 2A,B belongs to monoclinic crystal structure with the following dimensions; a = 7.095(12) Å, b = 11.374(19) Å, c = 11.66(2) Å and the angle of α = 90°, β 100.46(3) ° and γ = 90° for compound **1** that was reported synthetically in literature [11] and a = 10.1687(6) Å, b = 6.9037(5) Å, c = 21.3367(13) Å with the angle of β = 107.668(3), and α = γ = 90° for compound **2** that was documented elsewhere [26,27]. In the Supplementary Data (Table S1), full information on crystal structures of compounds **1** and **2** was provided. Optimization of isolated compounds **1** and **2** were performed using B3LYP-GD3 with theory level 6–311G (d,p) and optimized structure of both were displayed in Figure 2B and Figure 3B with numbering pattern of the investigated compounds. Geometric variables such as bond lengths, bond angle, and torsional angle obtained from optimized structures (**1** and **2**) by quantum chemical approach are shown in Table 1. The energy of optimized structures **1** and **2** has the energy of −1178.55087987 and −688.22154646 au, respectively. Compound **2** has lower energy (plus the negative) compared to compound **1**. It can also be assumed that compound **2** is more stable than compound **1**. In the optimized structure of 4,6-diacetylresorcinol (**1**), the bond lengths of C=O, C-O- and -OH were observed as 1.23, 1.24 and 1.05 Å, respectively, as given in Table 1 and these values were correlated to bond lengths obtained from X-ray single crystallography experiment (Table 1). Similarly, in the optimized of 3-O-methylellagic acid (**2**), the bond lengths of C12-O1, C13-O2, C17-O4, C24=O18, C24-O11, C25-O11, O2-H3, and C13-O1 were noticed as 1.34, 1.35, 1.19, 1.20, 1.39. 1.37, 0.96, and 1.43 Å, respectively, which displayed the presence of carboxylic, ester, and hydroxyl group in the structure of compound **2**. These bond length values were found to correspond with the values obtained as shown in Table 1. from a single X-ray study. Bond lengths and angles of certain atoms were found to marginally deviate from the corresponding experimental data. The deviation between certain atoms of the isolated compounds was found to be slightly large in the case of dihedral angles (°). Due to the non-covalent intermolecular interaction present in the crystal, these deviations may result from the fact that the theoretical calculations were performed under vacuum for a single isolated compound. The molecular geometries of the isolated compounds **1** and **2** obtained using XRD and DFT computation with basis set B3LYP/6-311G(d,p) were superimposed atom by atom (Figure 2C and Figure 3C), and RMSE values of both compounds were found to be 0.1345 and 0.1063 Å, respectively. These low values indicated that theoretical and experimental structural parameters agreed well.

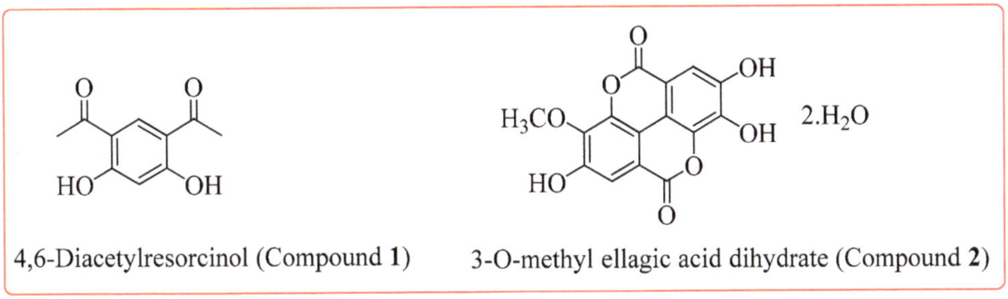

4,6-Diacetylresorcinol (Compound 1)   3-O-methyl ellagic acid dihydrate (Compound 2)

Figure 1. Isolated compounds (**1** and **2**) from *Bixa orellana*. leaves.

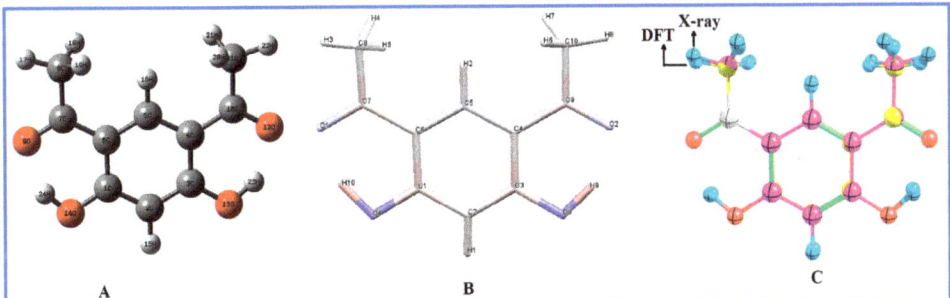

Figure 2. (**A**) Optimized geometry, (**B**) XRD of 4,6-diacetylresorcinol (**1**) and (**C**) superposition of the optimized structure on the experimental structure atom by atom.

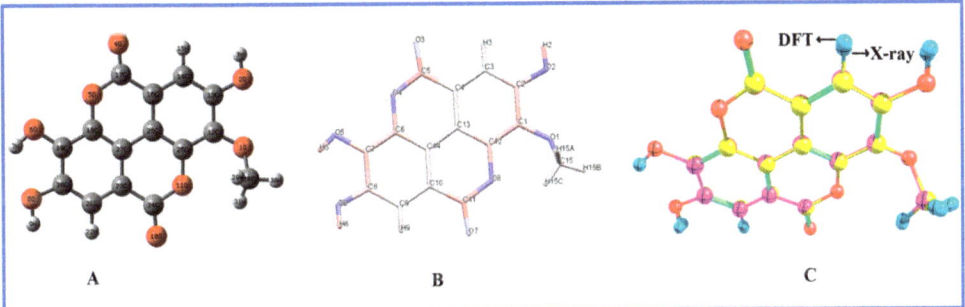

Figure 3. (**A**) Optimized geometry (**B**) XRD of 3-O-methylellagic acid dihydrate (**2**) and (**C**) superposition of the optimized structure on the experimental structure atom by atom.

Table 1. Comparison of important bond lengths [Å] bond angles [°] and dihedral angles [°] obtained by X-ray study and DFT/B3LYP/6-311G(d,p) for molecules **1** and **2**.

Bonds (Å)	Expt.	DFT	Angles (°)	Expt.	DFT	Dihedral Angles (°)	Expt.	DFT
				4,6-Diacetylresorcinol (1)				
C1-C2	1.38	1.39	C2-C1-C6	121.0	120.2	C6-C1-C2-H15	−175.2	−179.9
C1-C6	1.41	1.43	C2-C1-O14	117.9	118.1	O14-C1-C2-C3	179.0	180.0
C1-O14	1.34	1.33	C6-C1-O14	121.0	121.6	O14-C1-C6-C5	−179.5	−180.0
C2-C3	1.37	1.39	C1-C2-C3	119.8	120.8	O14-C1-C6-C7	−0.916	−0.004
C2-H15	0.99	1.08	C1-C2-H15	116.9	119.5	C1-C2-C3-O13	−178.7	−179.9
C3-C4	1.42	1.43	C3-C2-H15	122.8	119.5	C15-C2-C3-O13	−5.6	−0.0143
C3-O13	1.34	1.33	C3-C4-C4	117.9	120.2	C2-C3-O13-H23	174.6	179.9
C4-C5	1.38	1.39	C2-C3-O13	118.1	118.1	C4-C3-O13-H23	−5.0	−0.0051
C4-C10	1.46	1.46	C4-C3-O13	120.9	121.6	C3-C4-C10-O12	1.3	0.00076
C5-C6	1.39	1.39	C3-C4-C10	120.0	119.7	C5-C4-C10-O12	−178.2	179.9
C5-H16	0.99	1.08	C5-C6-C7	121.7	122.4	C5-C6-C7-O9	176.9	179.9
C6-C7	1.46	1.46	C6-C7-C8	120.9	120.0	O9-C7-C8-H17	10.5	0.0172
C7-C8	1.48	1.51	C6-C7-O9	119.8	121.0	O9-C7-C8-H18	−106.3	−120.0
C7-O9	1.23	1.23	C8-C7-O9	119.1	118.8	O9-C7-C8-H19	135.7	120.0
O14-H24	1.05	0.99	C1-C14-O24	101.6	106.9	O12-C10-C11-H20	−125.3	−119.9
				3-O-Methylellagic acid (2)				
O1-C12	1.35	1.34	C12-O1-C28	115.9	120.4	C28-O1-C12-C13	115.1	146.0
O1-C28	1.44	1.43	H3-O2-C13	109.5	108.9	C28-O1-C12-C25	−68.7	−36.5
O2-H3	0.82	0.96	C17-O5-C18	122.1	122.7	C12-O1-C28-H29	−53.0	−42.5
O2-C13	1.34	1.35	H7-O6-C19	109.4	108.0	H3-O2-C13-C12	172.8	179.2
O4-C17	1.20	1.19	C24-O11-C25	122.4	123.6	H3-O2-C13-C14	−7.3	−1.1
O5-C17	1.37	1.39	O1-C12-C13	118.7	116.0	C18-O5-C17-O4	−178.0	179.5
O5-C18	1.38	1.36	O1-C12-C25	122.0	126.0	C18-O5-C17-C16	1.9	−0.4
O6-H7	0.81	0.96	C13-C12-C25	119.0	117.7	C17-O5-C18-C19	178.1	−179.3
O6-C19	1.33	1.34	O2-C13-C12	114.8	115.7	H7-O6-C19-C18	−170.3	179.8
C18-C27	1.38	1.39	O2-C13-C14	124.3	122.9	H7-O6-C19-C20	10.2	−0.1
C23-C24	1.45	1.46	C13-C14-H15	120.4	120.8	O10-C24-O11-C25	178.9	−178.8
C19-C20	1.40	1.41	C13-C14-C16	119.1	120.0	O10-C24-C23-C21	0.9	−0.4
C28-H31	0.95	1.09	O8-C20-C19	113.9	113.6	H15-C14-C16-C17	−0.04	0.2
C21-H22	0.93	1.08	C19-C18-O5	118.3	118.4	H15-C14-C16-C26	179.4	−179.6
C16-C25	1.39	1.40	C18-C27-C26	118.7	118.5	C18-C27-C23-C24	178.5	179.6

The IR frequencies of the optimized geometry of the isolated compounds **1** and **2** were computed at the same basis set using a double harmonic approximation. Experimental and simulated IR spectra of compounds **1** and **2** are demonstrated in Figures 4 and 5. Table 2 displays the positions of some significant infrared (FTIR) vibrational bands of both compounds as well as their theoretical values (in cm^{-1}), intensities, and assignments. In general, harmonic frequencies overestimate the experimental frequency due to systematic errors triggered by basis set incompleteness error, different approximations, and lack of vibrational anharmonicity used in both compounds in the current DFT calculations. The IR data of both compounds **1** and **2** obtained from theoretical and experimental were compared with the literature [11].

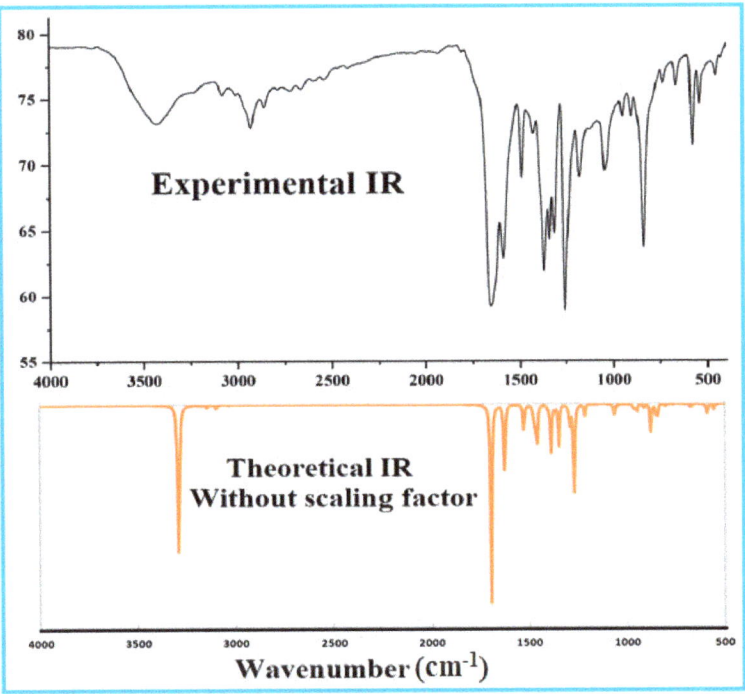

Figure 4. Theoretical and experimental IR spectra of compound 1.

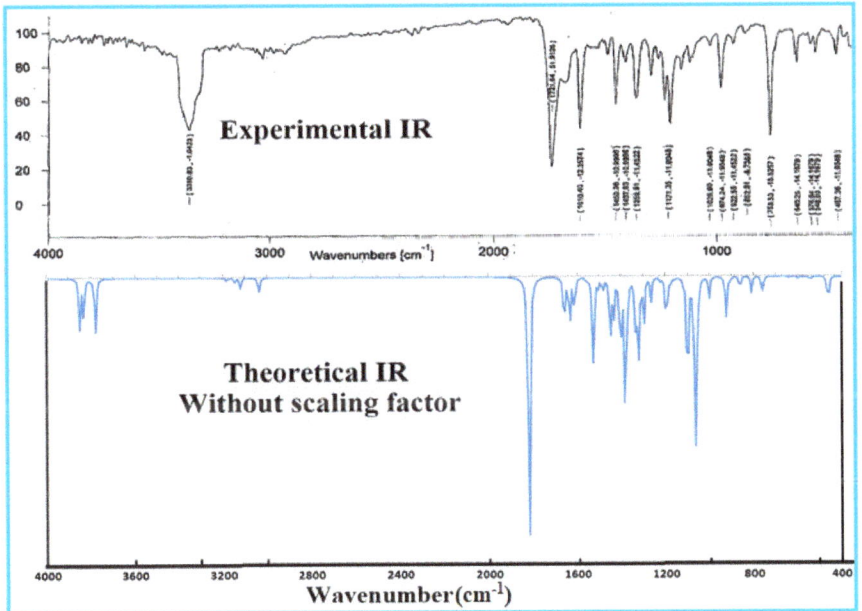

Figure 5. Theoretical and experimental IR spectra of compound 2.

Table 2. Experimental and theoretical IR frequencies (in cm^{-1}) along with IR intensity (in km/mol) and vibrational assignments of compound **1** and **2**.

	FTIR	$\omega_{harmonic}$	Assignments		FTIR	$\omega_{harmonic}$	Assignments
4,6-Diacetylresorcinol (1)	3427	3295	νOH	**3-O-Methylellagic acid dihydrate (2)**	3433	3848–3778	νOH
		3291	ν$_{asy}$OH		3159	3184–3179	νCH (ar)
		3212	νCH (ar)		2927	3146, 3119, 3035	ν$_{asy}$CH, νCH
	3082	3207	νCH (ar)		1723	1820, 1813	νC=O
		3147	νCH (ar)		1600	1659	νCC, βHOC
		3146	νCH (ar)			1650, 1634, 1612, 1560, 1549, 1526	νCC, ν$_{asy}$CC, γHCC
	2925 2851	3097	νCH		1518	1504	βCOC, βHCH, τHCOC
		3036	νCH		1422	1489	βHCH, τHCOC
		1696	ν$_{asy}$C=O			1450	νCC, βHOC
	1652	1694	νC=O		1361	1382	νO-C, ν$_{asy}$C-O
		1661	ν$_{asy}$C=O, Ar-ring		1340	1372, 1322	ν$_{asy}$CC, βHOC
	1588	1627	νCC, ν$_{asy}$CC, βHOC		1293	1319	ν$_{asy}$O-C, βHCC
		1527	βHOC			1297	νO-C, γHOC, βHCC
	1486	1482	βHCH, τHCCC		1190	1221	βHOC, βHCH, τHCOC
		1472	βHCH, τHCCC			1175	βHOC, βHCC
	1425	1398	βHCH		1060	1167	βHCH, τHCOC
		1290	νO-C and βHCC (ar)		966	1053, 1000	νO-C
		1266	νCC, βHOC, βCCC		912	921	γCCC
		1214	νO-C, βHCC (ar)		867	861	τHCCC
		1092	βCCC, τHCCC			859	outCCCC
		964	γHCH, τHCCC		800	807	mix vibration
		914	τHCCC			759	outOCOC
		762	βCCC			699	βCOC, outOCCC
		725	outCCCC			640	βCOC, βOCC,
		674	βCCC		575	614	βCOC, βOCC
		556	βCCC, βCCO			539	νCC, βCCC,

ν-stretching, ν$_{asy}$ asymmetric stretching, β-in plane bending, γ-out of plane bending vibrations, t-torsional vibration.

Both compounds **1** and **2** have groups of C-H, aromatic, OH, and C=O in their structure. Simulated IR spectra using B3LYP/6-311G(d,p) basis set were compared with experimental IR spectra to explain vibrations of functional groups, as shown in Figures 4 and 5. All assignments set out in Table 2 are in accordance with the literature [28]. The bands due to stretching C-H, C=O, C=C, C-C, ring, C-O as well as in-plane and out-of-plane deformation vibrations are shown in Table 2. The bands that appeared in the region of the lower wavenumber were mainly due to mixed vibrations of torsion and out-of-plane deformation of the rings. In general, the IR bands vibrate in the range of 3700–2700 cm^{-1} due to OH groups (IR Spectrum Table and Chart available on Sigma Aldrich website), with different intensities showing the nature of –OH groups such as free, inter, or intramolecular bonding in chemical compounds. The IR bands displayed in a broad peak in compound **1** and **2** at 3427 and 3433 cm^{-1}, respectively in IR spectra, assigning OH groups to hydrogen bonding interactions. In the theoretical spectra of 4,6-Diacetylresorcinol (**1**) and 3-O-methylellagic acid dihydrate (**2**), phenolic hydroxyl groups were vibrated at 3295–3291cm^{-1} and 3848–3778 cm^{-1}, respectively, without applying the scale factor. In the aromatic structure, the characteristic region for C-H stretching bands usually falls just above 3000 cm^{-1}. The stretching bands occurred in multiple weak and narrow peaks at 3082 cm^{-1} for **1** and 3159 cm^{-1} for compound **2** of the FTIR spectra while theoretical values of aromatic C-H were observed at 3212–3146 cm^{-1} and 3184–3179 cm^{-1}, respectively. In-plane and out-of-plane C-H bending vibrations exhibit their characteristic bands in the region 1100–1500, 800–1000, and 650–1000 cm^{-1}. In present compounds **1** and **2** in-plane deformation vibrations were observed mixing with vibrations of C-C, C-O-C stretching bands at 1482, 1472, 1290, and 1214 cm^{-1} for **1** and at 1489, 1319, 1297, 1221, 1175, and 1167 cm^{-1} for compound **2**, while out of plane bending vibration were noticed mixing with torsional and out vibration as shown in table. The asymmetric and symmetric stretching bands of the methyl group (C-H) of chemical compounds in FTIR have been documented below 3000 cm^{-1} in the 3000–2840 cm^{-1} range. FTIR spectra of both compounds exhibited bands at 2925, 2851, and 2927 cm^{-1}, representing asymmetric and symmetric stretching bands suggested –CH$_3$ groups in both compounds **1** and **2**, while the calculated values at 3097, 3146, 3119, and 3035 cm^{-1} (without scaling factor) were assigned to methyl groups in the present compounds. In general, the stretching vibration of the normal carbonyl group referred to without hindrance environment appears with a high intensity peak of about 1700 cm^{-1}. Strong bands visualized at 1652 and 1721 cm^{-1} with overlapping shoulders were assigned to C=O stretching vibrations while calculated bands of carbonyl groups using B3LYP/6–311G(d,p) basis set in IR spectra were identified at 1696, 1694, 1661 cm^{-1} for **1** and 1820, 1813 cm^{-1} for **2**. In most cases, the C-O stretching vibration of the aromatic ether, alkyl aryl ether, tertiary alcohol, ester, aliphatic ether, secondary alcohol, and primary alcohol attached to the heterocyclic moiety can cause a strong to medium intensity band in the 1310–850 cm^{-1} region. In the present study, stretching vibration bands at 1425, 1361, and 1293 cm^{-1} were observed in FTIR spectra of compounds (**1** and **2**). Bands at 1290, 1214, 1382, 1319, 1297, 1053, and 100 cm^{-1} may be due to the various nature C-O stretching vibrations with mixing of other vibrations in the theoretical study, as shown in table. The carbon-carbon stretching vibrations of aromatic ring crop up in the region 1625–1430 cm^{-1}. The C=C stretching vibration of aromatic ring present in compounds (**1** and **2**) display their characteristics bands at 1588 and 1425 for **1**, 1600, 1422 and 1293 for **2** in FTIR while calculated bands at 1627 and 1266 for **1**, 1659, 1450 and 1297 for **2** were noted with mixing of other vibrations of various intensity in IR spectra of the isolated compounds.

3.2. Frontier Molecular Orbital and UV-Visible Spectral Analysis

Frontier molecular orbitals (FMOs) referred to HOMO and LUMO involved in chemical reactions. Figure 6 displays the highest occupied molecular orbitals (HOMO) and lowest unoccupied molecular orbital (LUMO) for compounds **1** and **2** isolated from plant. HOMO is the orbital, which primarily acts as an electron donor, and LUMO is the orbital, which primarily acts as an electron acceptor, and molecular chemical stability is defined by the difference between HOMO and LUMO. It has been noted that the highest occupied molecular orbitals (HOMOs) are largely distributed over aromatic rings, including C=O and –OH functional groups excluding C2H15, C5H16 and methyls in compound **1** and O5 in compound **2** whereas LUMOs is completely delocalized on all of the molecules except OH groups in compound **1** and two OH groups in compound **2** as seen in Figure 6. The energy gap (Eg) between HOMO and LUMO is a critical parameter to study the nature of chemical reactivity because it is a parameter of electron transfer from HOMO to LUMO.

The large value of the gap specifies the most stability and the least reactivity and *vice versa*. The energy gaps of compounds **1** and **2** were found to be 4.73 and 3.97 eV, respectively. The low energy gaps of the two compounds provide an indication of the efficient electronic transition as well as their high reactivity. The HOMO energy describes the susceptibility of the molecule towards electrophilic attacks, while the LUMO energy represents the susceptibility of the molecule to nucleophilic attacks. The lowest singlet to singlet spin allowed transitions, oscillator strengths, absorption wavelengths, and excitation energies for 4,6-diacetylresorcinol (**1**) and 3-*O*-methylellagic acid dihydrate (**2**) in solution phase and comparison of experimental and theoretical UV-vis absorption spectra of both compounds **1** and **2** are presented in Table 3 and Figures 7 and 8. The UV-vis bands observed at 272 and 323 for compound **1** and 255 and 365 nm for compound **2** in solution were assigned to H→L + 1, H-1→L + 1, H→L, and H-1→L + 1, H→L, H-3→L + 1, H-2→L transitions for **1** and H-3→L, H-5→L, H-4→L and H→L, H-1→L, H→L + 1 transitions for compound **2** as shown in Table 3. The corresponding calculated wavelengths were observed to be 268.5, 251.9, 302.6, and 300.1 nm for compound **1** and 259.7, 252.9, 351.7, and 313.6 nm for compound **2** in solution. The alteration of the predicted wavelengths was due to the solvent effect as observed in the two compounds (**1** and **2**).

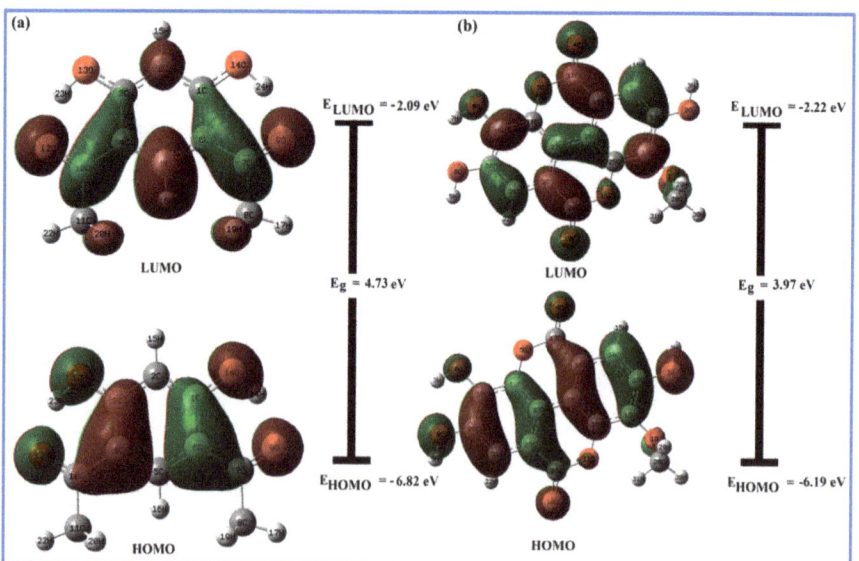

Figure 6. Energy levels, gaps and localization distribution of HOMO and LUMO (frontier molecular orbitals) of the (**a**,**b**) for compounds **1** and **2**, (energies are given in eV).

Table 3. UV/Vis spectral characteristic of compounds **1** and **2** in the solution phase.

	λ_{exp}	λ_{theo}	E(eV)	f	Composition (%)		λ_{exp}	λ_{theo}	E(eV)	f	Composition (%)
4,6-Diacetylresorcinol (1)	273.6	268.5	4.6167	0.1915	H→L + 1 (99)	3-O-Methylellagic acid dihydrate (2)	255.6	259.7	4.77	0.0498	H-3→L(30)
		251.9	4.9212	0.7777	H-1→L + 1 (69), H→L (25)			252.9	4.90	0.0366	H-5→L (25), H-4→L (50)
	323.7	302.6	4.0968	0.1209	H-1→L + 1 (26), H→L (73)		365.3	351.7	3.52	0.2290	H→L(94)
		300.1	4.1314	0.0366	H-3→L + 1 (19), H-2→L (79)			313.6	3.95	0.0269	H-1→L (51), H→L + 1 (4)

Abbreviation used: H-HOMO, L-LUMO, λ-wavelength (in nm), E-excitation energy, f-oscillator strength.

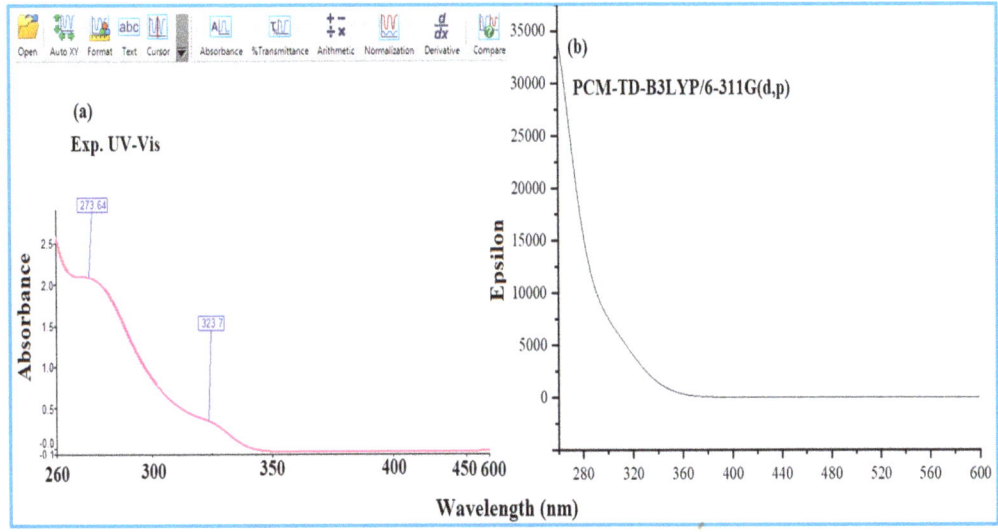

Figure 7. Experimental (**a**) and calculated (**b**) UV-vis absorption spectra of the 4,6-Diacetylresorcinol.

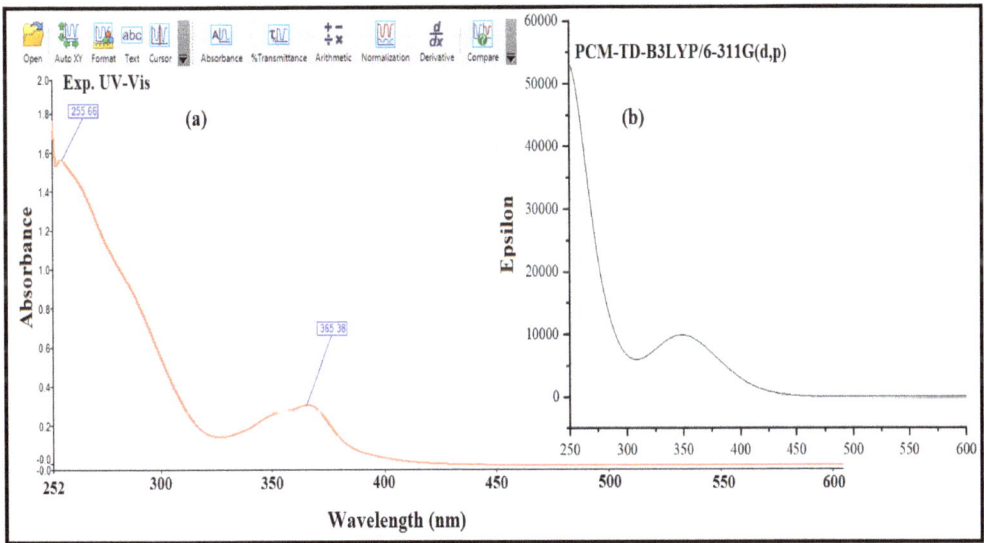

Figure 8. Experimental (**a**) and calculated (**b**) UV-vis absorption spectra of the 3-O-methylellagic acid dihydrate.

3.3. NMR Spectral Analysis

The ^{1}H and ^{13}C-NMR spectra of compounds (**1** and **2**) and their chemical shifts were given in ppm downfield form of tetramethylsilane (TMS) and illustrated in Figures 9–12. Figure 9 & Figures 10 and 11 & Figure 12 depict the ^{1}H and ^{13}C NMR spectra of compounds 1 and 2, respectively. Theoretical calculations were performed to validate the accuracy of the experimental data collected. Prior to calculating the theoretical chemical shifts in term of NMR, the molecular geometry of the compounds was optimized using GIAO-B3LYP with 6–311G(d,p) level of theory. Deuterium exchangeable proton of the –OH group in **1** showed a chemical shift at 129 ppm based on its integration of two protons as a singlet peak in ^{1}H-NMR spectrum of compound 1. Its corresponding calculated chemical shift was observed at 12.87 ppm. Other singlets resonated at 8.2 and 6.4 ppm showed aromatic protons flanked by methyl ketone and hydroxyl groups, respectively, whereas theoretical singlets of aromatic protons were resonated at 8.1 and 6.3 ppm in solution phase suggesting that chemical shifts were quite similar to experimental values leading to the identification of exact structure of compound **1**. Chemical shifts at 2.1 and 1.6 ppm may contribute to the methyls attached to the ketone groups, corresponding singlets at 2.7 and 1.9 ppm in theoretical calculations were very close to experimental studies as shown in Figure 9. As shown in Figure 11 of compound **2**, singlets refer to methyl hydrogens divided into three and observed in the theoretical spectrum at 3.5, 4.1 and 4.8 ppm, which were observed as a peak at 4.2 ppm in experimental spectrum because the methyl protons experienced the same chemical environment that caused their overlaps. At 7.4 and 7.6 ppm, aromatic protons were noted as singlets of two aromatic ring protons and corresponding peaks observed in the theoretical spectrum at 7.60 and 7.63 ppm were very close to the experimental spectrum. The deuterium exchangeable proton of the three –OH groups in compound **2** displayed chemical shifts at 4.8, 5.0, and 6.0 ppm as three singlet peaks in theoretical ^{1}H NMR spectrum of the compound **2**. However phenolic groups attached to various chemical environments on aromatic rings were observed at 8.7 ppm for one proton and 10.5 ppm assigned to two protons. The theoretical values of NMR displayed good correlation with experimental values for compounds **1** and **2** except discrepancy was noted for phenolic protons because of lability of the hydroxyl proton and proximity of another phenolic group for compound **2**.

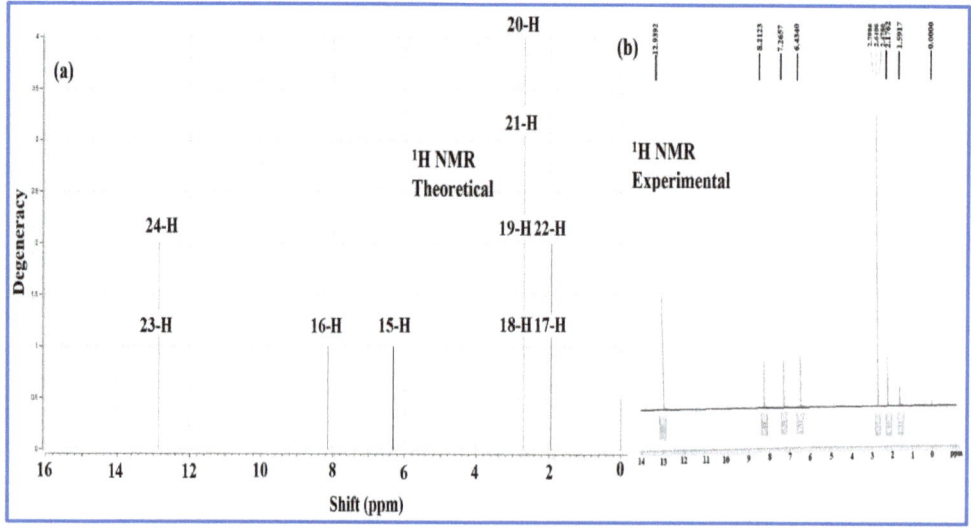

Figure 9. Calculated (**a**) and experimental (**b**) ^{1}H-NMR spectra of the isolated compound **1**.

The signals characterized in the ^{13}C NMR spectra of molecules **1** and **2** (Figures 10 and 12) showed resonance at 206 and 175 for **1** and 162, 162, 159, 154.4, and 144.1 ppm for compound **2**, indicating the presence of carbonyl carbons and carbons attached –OH groups present in aromatic rings in various chemical environments with different chemical shifts in isolated compounds **1** and **2**. Corresponding signals in experimental ^{13}C NMR of compounds **1** and **2** were observed at 202 and 168.9 ppm for **1** and 159, 154, and 143.2 ppm for **2** are closed to theoretical ^{13}C NMR spectra of both compounds. Other important signals observed in the theoretical ^{13}CNMR spectrum at 106.4, 115.3, and 141 ppm were attributed to carbon flanked by –OH groups, carbons attached to carbonyl groups, and carbon flanked by carbonyl groups for compound **1** were found to correlate with the signals at 105, 113.6, and 136.2 ppm in experimental ^{13}C NMR of compound **1**. Signals resonated at 26 ppm was attributable to carbon atom of the methyl groups bonded to the oxygen atom in experimental spectrum of compound 1. The estimated theoretical value of 26.4 ppm of methyl carbon is in accordance with the experimental results. Some other ^{13}C NMR signals of aromatic carbons directly influenced by functional groups and substituted carbon atoms including methyl carbon were observed in theoretical spectrum at 140 (-OH-C-C18-O-), 146 (-C12-O-CH3) and 149 ppm (C25-O-C=O), 112.3 (H-C21), 114.2 ppm (H-C14) and 61.2 ppm (O-C28-) that were resonated in similar pattern with slightly deviation in experiment ^{13}C NMR of compound 2.

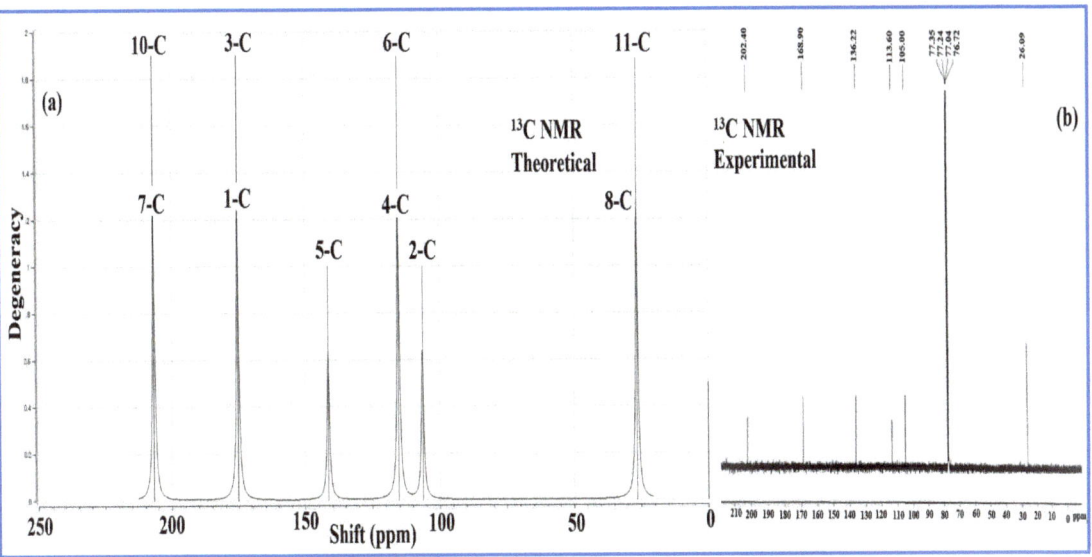

Figure 10. Calculated (**a**) and experimental (**b**) ^{13}C-NMR spectra of the isolated compound 1.

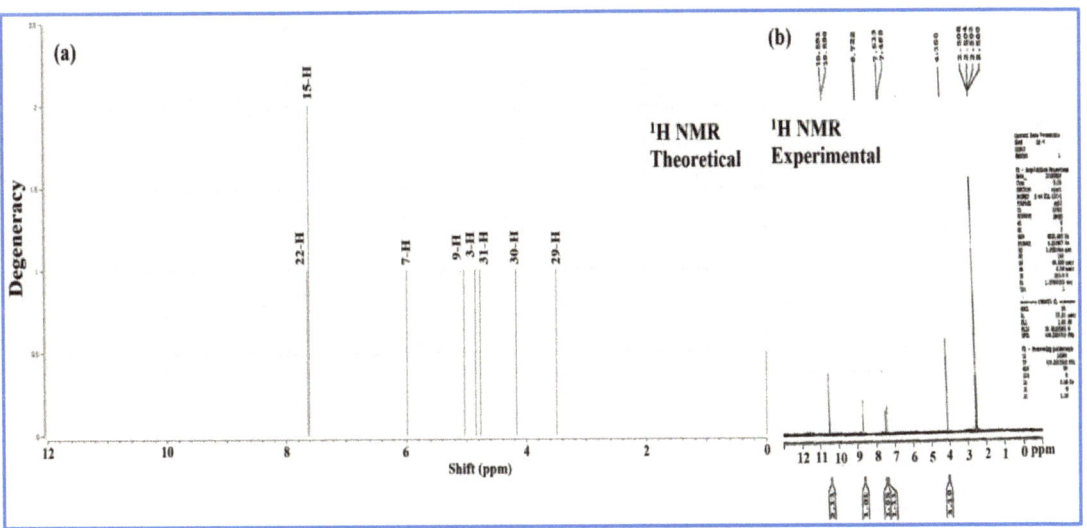

Figure 11. Calculated (**a**) and experimental (**b**) ^{1}H-NMR spectra of the isolated compound 2.

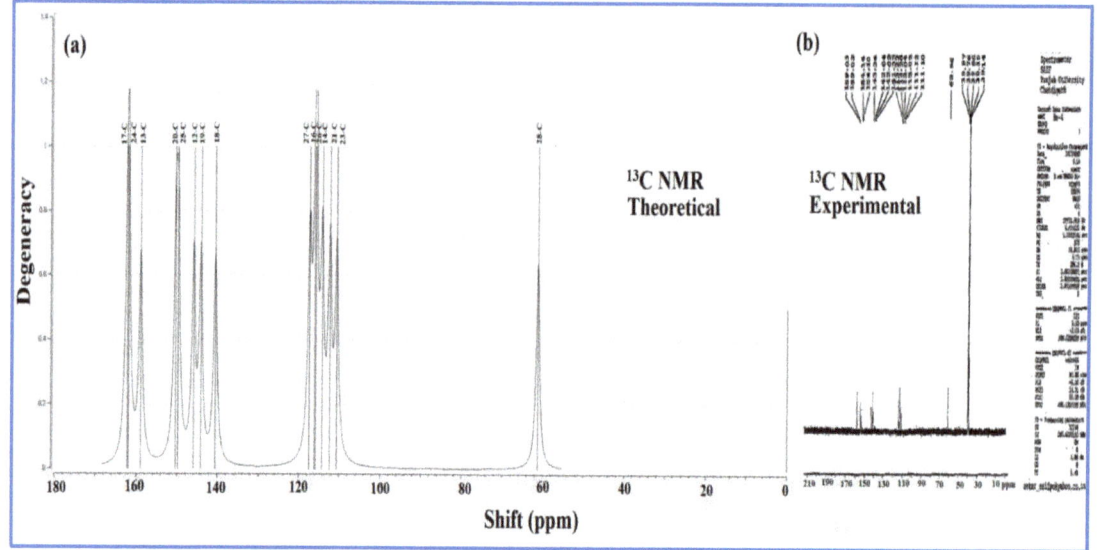

Figure 12. Calculated (**a**) and experimental (**b**) ^{13}C-NMR spectra of the isolated compound **2**.

3.4. Molecular Electrostatic Potential (MEP)

In the study of biological recognition processes and hydrogen bonding interactions, the MEP map is a useful tool for qualitatively interpreting electrophilic and nucleophilic reactions [29–31]. Molecular electrostatic potential of 4,6-diacetylresorcinol (**1**) and 3-O-methylellagic acid dihydrate (**2**) are calculated using the optimized structures of both compounds applying at B3LYP/6-311G(d,p) basis set, and their plots are shown in Figure 13. Colored figures of compounds **1** and **2** were a significant sign of inter- and intramolecular interactions and reactivity of molecules. The red to blue color of the graphs reflects the electron-rich to electron-poor regions. Carbonyl oxygen and hydroxyl oxygen atoms (O13, O14, for **1** and O2, O6, O8 with ionizable protons) at the bottom of the reddish region have an electron rich area of **1** and **2**, potentially the most aggressive nucleophilic attack region. On the other hand, an electron poor region with a bluish color overlying a phenolic proton of the isolated compounds that may demonstrate an electrophilic behavior.

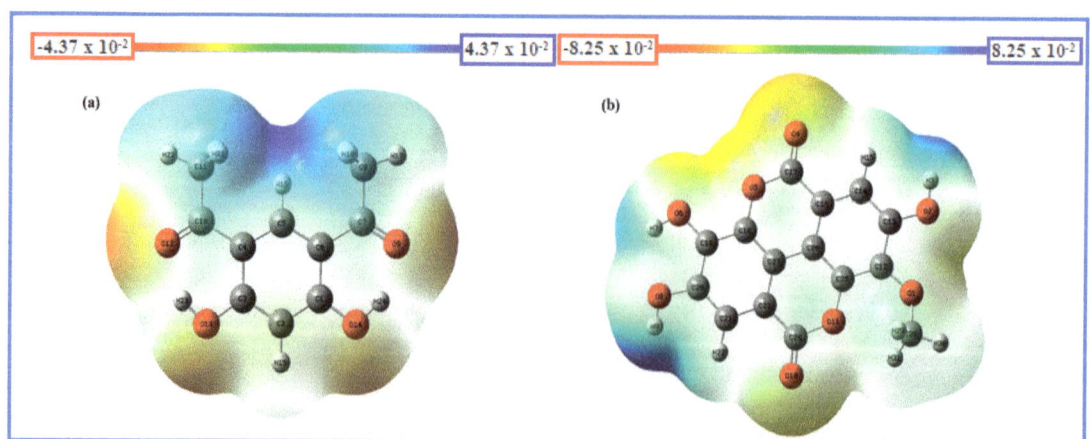

Figure 13. Molecular electrostatic potential surface of the compound **1** (**a**) and compound **2** (**b**) having regions rich and poor in electrons, as indicated by the colors red and blue.

3.5. UV-Visible Absorption Spectroscopy

UV–Visible absorption spectroscopy is one of the most fundamental and extensively used experimental techniques for evaluating the stability of ct-DNA and interactions with small ligand molecules [32]. Complex formation is usually accompanied by changes in the intensity and position of the absorption spectra [33]. The absorption maximum of compound **2** was centered on 273 nm. The maximum absorption band of compound **2** increased with increasing ct-DNA concentrations, accompanied by a shift of 11 nm, indicating the formation of a complex, as illustrated in Figure 14. Intercalation as a possible method of binding was discarded because it typically involves hypochromic or isosbestic points [32].

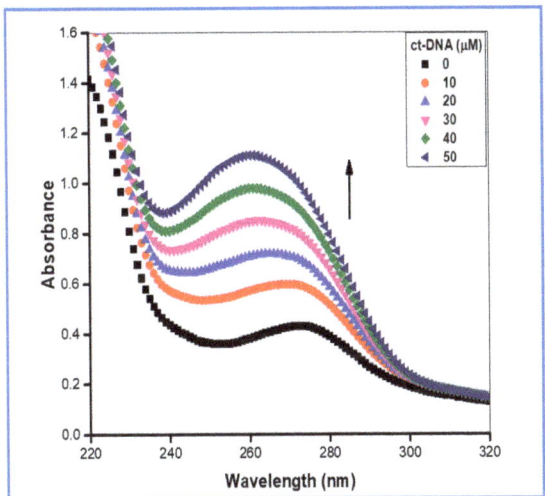

Figure 14. UV-absorption spectra of compound **2** (200 µM) in the presence of increasing concentration of ct-DNA (0–50 µM).

3.6. Steady-State Fluorescence

Fluorescence spectroscopy is undoubtedly one of the extensively used techniques for studying the interactions between small ligand molecules and DNA. High sensitivity, large linear concentration range, and selectivity are its advantages over other techniques. It can also be used to determine the pattern of compound binding to ct-DNA, as well as providing a variety of other details about the binding mode, strength, and number of binding sites in ct-DNA [34]. Compound **2** (200 µM) was excited at 273 nm and showed emission maxima at 367 nm. As demonstrated in Figure 15a, after consecutive additions of ct-DNA (0–125 µM), the intensity of compound **2** emission grew dramatically. This suggests the possibility of a complex-forming between them [35].

The Stern–Volmer equation [36] was used to estimate the enhancing mechanism involved in the process of binding between ct-DNA and compound **2**:

$$\frac{F_0}{F} = 1 - K_{SV}[Q] \tag{1}$$

$$K_q = \frac{K_{SV}}{\tau_0} \tag{2}$$

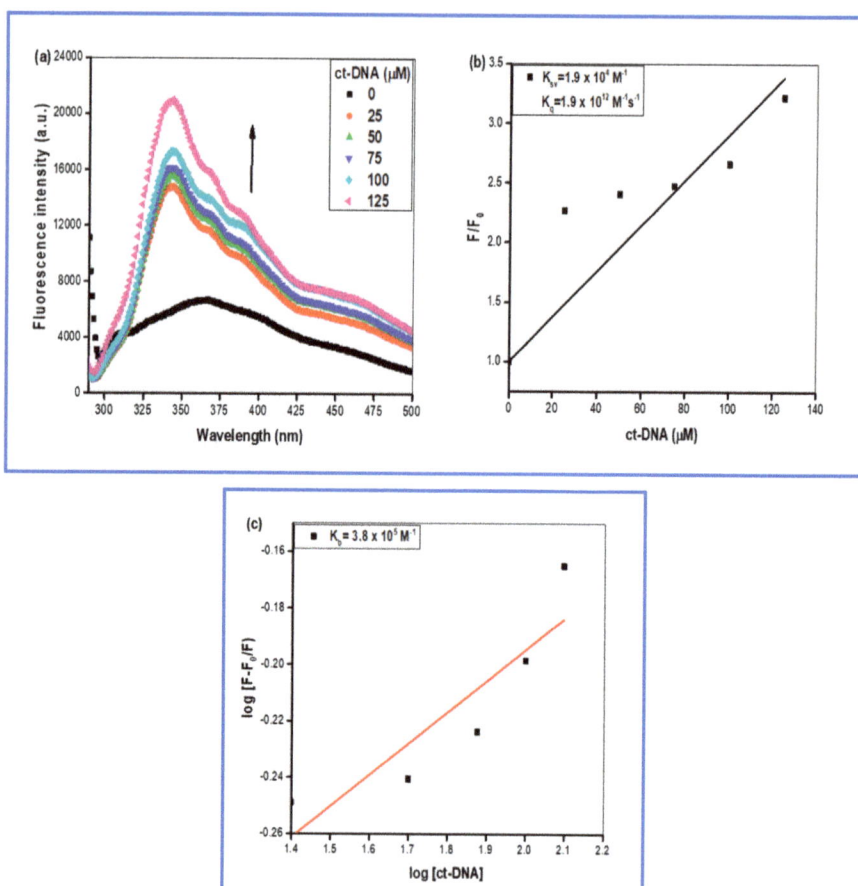

Figure 15. (a) Steady-state fluorescence spectra of compound **2** in the presence of increasing concentration of ct-DNA (0–125 μM). (b) Stern–Volmer plot of compound **2** complexed with ct-DNA to calculate K_{sv} and K_q. (c) Double logarithmic regression plot to calculate the binding constant (K_b).

The fluorescence intensities in the absence and presence of the quencher (ct-DNA) are represented by F_0 and F, respectively. The Stern–Volmer and the bimolecular enhancement constant are represented as K_{sv} and K_q, respectively. In the absence of the quencher, τ_0 is the fluorophore's average lifespan ($\tau_0 \approx 10^{-8}$ s) [37].

The binding process can be improved in one of two ways: by using a static or dynamic method. Static enhancement involves the formation of a ground state complex [38], while dynamic enhancement is largely associated with molecular diffusion.

The slope of the Stern–Volmer plot was used to obtain the K_{sv} value for compound **2** in Figure 15b. Equation (2) was used to calculate K_q using the value of K_{sv}, and both results are given in Table 1. Since the values of K_q exceeds the threshold value for dynamic enhancement 2×10^{11} M^{-1} s^{-1}, this suggests that rather than a dynamic approach [32], a static improvement technique should be used.

The binding constant (K_b) and interaction stoichiometry (n) were calculated using a double logarithmic regression plot. [32]:

$$\frac{\log(F - F_0)}{F} = \log K_b + n \log[\text{DNA}] \tag{3}$$

where K_b is the binding constant, and n is the number of binding sites of compound **2** in ct-DNA double helix. As shown in Figure 15c, the log $[(F - F_0)/F]$ vs. log [ct-DNA] plot's slope and intercept were calculated using Equation (3).

The value of n was discovered to be **1** when compound **2** was complexed with the ct-DNA interaction. As a result, compound **2** appears to have only one binding site in ct-DNA. The values of K_b for compound **2** were also determined and are presented in Table 4.

Table 4. From fluorescence measurements, values of Stern–Vomer, bimolecular enhancement constant, binding site number and binding constant achieved.

System	K_{sv} (M^{-1})	K_q (M^{-1} s^{-1})	n	K_b (M^{-1})
Compound 2	1.9×10^4	1.9×10^{12}	1.0	3.8×10^5

3.7. Circular Dichroism (CD)

Modifications in the secondary structure of ct-DNA [35] can be tracked using a useful tool, CD spectroscopy. By monitoring variations in the CD spectra of ct-DNA, binding mechanism of compound **2** to ct-DNA was also confirmed. Base stacking interactions cause a negative peak at 245 nm, and helicity induces a positive peak at 275 nm in the CD spectra of ct-DNA [39]. Intercalation entails the creation of an intercalation cavity in which the compounds can bind, as well as the development of new molecular connections that anchor the complex. Groove binding, on the other hand, does not require such conformational changes [37]. Intercalative molecules impair the ct-DNA interaction with base stacking, altering the negative peak as a result. Groove binders have no effect on the structure of ct-DNA; hence no such modifications are apparent [35]. No substantial change in the CD spectra of ct-DNA (200 µM) was found when compound **2** (200 µM) was added, indicating a groove binding interaction as seen in Figure 16.

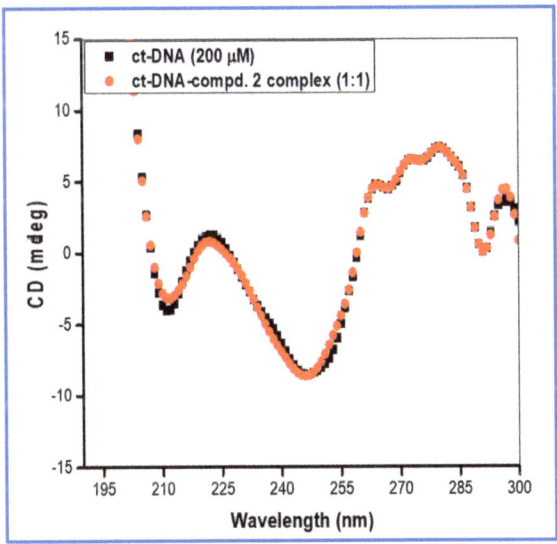

Figure 16. The CD spectra of ct-DNA (200 µM) alone and ct-DNA complexed with compound **2** in the molar ratio 1:1.

3.8. Docking

Molecular docking is a scoring function-based technique for predicting the interaction mechanism of a small molecule in a protein's binding site with the optimal orientation and affinity [36]. It is also utilized to back up the findings obtained through spectroscopy and fluorescence-based experiments. Compound **2** was docked against B-DNA dodecamer d(CGCGAATTCGCG)2 (PDB ID:1BNA). Docking studies reveal that molecule **2** binds in DNA's minor grooves with a binding energy of −8.7 kcal mol^{-1}, as illustrated in Figure 17a. Compound **2** engages with DNA via two hydrogen bonds (DG) with bond lengths of 2.34 and 2.31 Å, which help compound **2** binding to DNA to be stable.

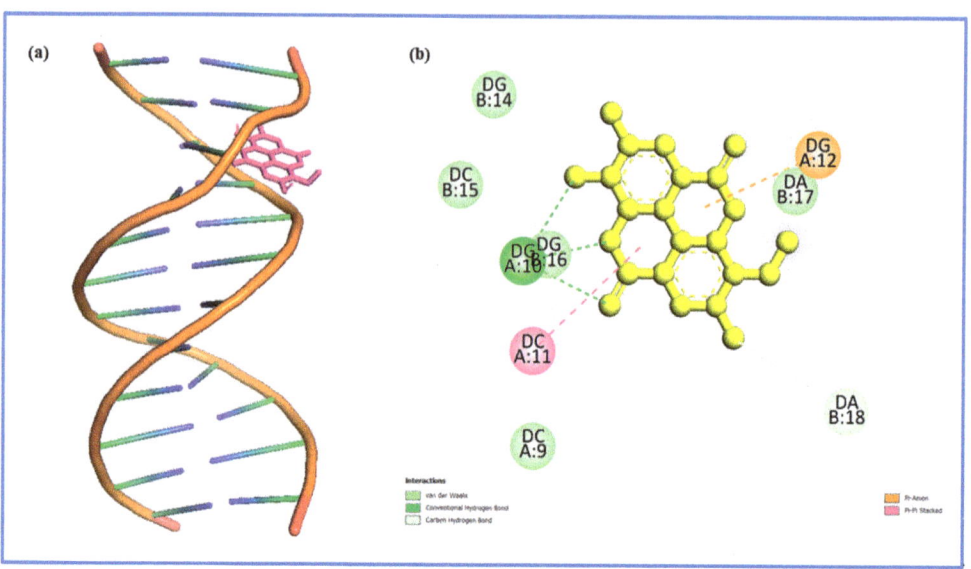

Figure 17. (**a**) Docked pose of compound **2** binding with DNA. (**b**) 2D-plot of the interaction.

3.9. Prediction of DRUG-likeness (Lipinski's Rule of Five)

The early preclinical analysis is greatly assisted by drug-likeness filters, which help to avoid costly late-stage preclinical and clinical failure. The drug-likeness properties of the compound were studied by using the Lipinski rule of five. It is a preliminary criterion for assessing its structural resemblance to an ideal drug [40,41]. Compound **2** under investigation follows all of the parameters under the 'Rule of Five', with no violations calculated via an online server, and revealed a higher tendency of compound **2** towards drug-likeness.

3.10. Bioavailability Score and Synthetic Accessibility

Bioavailability difficulties linked with a compound might slow down the therapeutic development process. For a compound to exhibit its pharmacological action on the body and be effective as a drug, it must be available in sufficient concentration in the systemic circulation for a specific duration [42]. To avoid unfavorable outcomes in the future, it is important to evaluate the compound's bioavailability early in the therapeutic development process. Therefore, the bioavailability of compound **2** was determined with the SwissADME web tool [25]. This web tool determines the bioavailability of the compound based on the molecular properties and lipophilicity by applying through different principles such as Lipinski's rule of five, Ghose filter, Veber filter, Egan filter, and Muegge filter. The bioavailability score of the isolated molecule (**2**) was found to be 0.55, which implies that there is a 55% probability of being bioavailable.

To identify a promising lead compound during the process of virtual screening, it is preferred to filter out a non-toxic, physiologically active compound with good bioavailability. In addition, the degree of complexity in synthesizing a compound is a factor that should be considered while choosing the most promising compound [43]. The SwissADME web tool was also used to measure the degree of difficulty in synthesizing compound **2**. Synthetic accessibility is a fingerprint-based computational approach for determining how difficult it is to synthesize a compound. A synthetic accessibility value of **1** indicates that the compound is relatively easy to synthesize, while the synthetic accessibility score of 10 indicates that the compound is extremely difficult to synthesize. Compound **2** was found to have a synthetic accessibility score of 3.21, indicating that it will be easier to synthesize.

4. Conclusions

In this paper, two compounds named as 4,6-Diacetylresorcinol (**1**) and 3-O-methylellagic acid dihydrate (**2**), were isolated from the leaves of *Bixa orellana*, although they have previously been described in other sources and synthetic methods. The density functional theory was performed to calculate the electronic structure in support of experimental results. In order to compare spectral data of experimental studies (IR, ^{1}H, ^{13}C NMR, UV and parameters obtained from single X-ray diffraction) with theoretical exploration, B3LYP/6-311G(d,p) was used to optimize geometric parameters in the gas phase, while the NMR and UV-Vis studies of the optimized compounds performed by GIAO-B3LYP/6-311G(d,p) and PCM-B3LYP/6-311G(d,p) level theory were applied in the respected solution phase, respectively. The theoretical and experimental observations of the isolated compound were in good agreement with the establishment of the chemical structures of both molecules. In addition, energy of HOMO, LUMO and its gap including the MEP for compound **1** was calculated that were found to be −6.82 eV, 2.09 eV, and 4.73 eV, respectively, while compound **2** was −6.19 eV, 2.22 eV, and 3.97 eV respectively, these energy gaps of 4,6-diacetylresorcinol and 3-O-methylellagic acid dihydrate, including MEP, were an indicator of successful electronic transition and their high reactivity including chemical nature of molecules having various functional groups for the future course of the drug development response. Using multiple biophysical and in-silico approaches, the present study adds to our understanding of the interaction of molecule 2 with ct-DNA. UV-absorption and fluorescence spectroscopy confirm the complex formation between compound **2** with ct-DNA. The value of binding constant (K_b) obtained was in the order of 10^5 M^{-1}. Circular dichroism results of ct-DNA complexed with compound **2** suggest groove binding mode. Furthermore, *in silico* studies also suggest that the binding of compound **2** occurs in the minor groove of the DNA. Lipinski's rule of five revealed a higher tendency of compound **2** towards drug-likeness. The bioavailability score and synthetic accessibility score of compound (**2**) was found to be 0.55 and 3.21, suggesting that compound **2** could serve as effective therapeutic candidate.

Supplementary Materials: The following supporting information can be downloaded at: https://www.mdpi.com/article/10.3390/cryst12030380/s1. Table S1. The experimental parameters related to the monocrystalline X-ray examination of **1** and **2**.

Author Contributions: Conceptualization, M.P., M.A. (Mohammad Azeem), A.M.M. and M.A. (Mahboob Alam); methodology, M.A. (Mahboob Alam), V.H.R. and A.A.; software, K.M., S.S., S.I.A.-R. and M.T.; validation, M.A. (Mahboob Alam) and A.A.; formal analysis, M.A. (Mohammad Azeem) and V.H.R.; investigation, M.A. (Mohammad Azam) and S.I.A.-R.; resources, M.A. (Mahboob Alam) and A.A.; writing—original draft preparation, M.A. (Mohammad Azam), S.I.A.-R., A.A. and K.M.; writing—review and editing, S.S., M.T. and M.P.; visualization, M.A. (Mahboob Alam); supervision, M.P., M.A. (Mohammad Azam) and K.M.; writing—original draft preparation, validation, K.M. All authors have read and agreed to the published version of the manuscript.

Funding: This research received no external funding.

Institutional Review Board Statement: Not applicable.

Informed Consent Statement: Not applicable.

Data Availability Statement: Not applicable.

Acknowledgments: The authors acknowledge the financial support through Researchers Supporting Project number (RSP-2021/147), King Saud University, Riyadh, Saudi Arabia. M. Azeem thanks Department of Chemistry, A.M.U, Aligarh, for providing the necessary research facilities. UGC is acknowledged for financial support. The authors acknowledge the Laboratory for Advanced Computing at University of Coimbra (http://www.uc.pt/lca) for providing computing resources that have contributed to the research results reported within this paper.

Conflicts of Interest: The authors declare no conflict of interest.

References

1. Silva, S.N.S.; Amaral, C.L.F.; Reboucas, T.N.H. Adoption of conservation practices on farm and selection of varieties by producers of annatto in the city of Vitoria da Conquista-BA. *Rev. Bras. Agroecologica* **2010**, *5*, 106.
2. Correa, M.P. *Dicionario das Plantas ´Uteis do Brasil e das ´ Exoticas Cultivadas*; Ministerio da Agricultura/IBDF: Rio ´de Janeiro, Brasil, 1978; Volume 4.
3. Villar, R.; Calleja, J.M.; Morales, C.; Caceres, A. Screening of 17 Guatemalan medicinal plants for platelet antiaggregant activity. *Phytother Res.* **1997**, *11*, 441. [CrossRef]
4. Aher, A.A.; Bairagi, S.M. Formulation and evaluation of herbal lipstick from colour pigments of *Bixa Orellana* (Bixaceae) seeds. *Int. J. Pharm. Biol. Sci.* **2012**, *4*, 357.
5. Kang, E.J.; Campbell, R.E.; Bastian, E.; Drake, M.A. Invited review: Annatto usage and bleaching in dairy foods. *J. Dairy Sci.* **2010**, *93*, 3891. [CrossRef] [PubMed]
6. Venugopalan, A.; Giridhar, P.; Ravishankar, G.A. Food, ethanobotanical and diversified applications of *Bixa orellana* L.: A scope for its improvement through biotechnological mediation. *Ind. J. Fundament. Appl. Life Sci.* **2011**, *1*, 9.
7. Scotter, M. The chemistry and analysis of annatto food colouring: A review. *Food Addit. Contam.* **2009**, *26*, 1123. [CrossRef]
8. Mercadante, A.Z.; Steck, A.; Pfander, H. Three minor carotenoids from annatto (*Bixa orellana*) seeds. *Phytochemistry* **1999**, *52*, 135–159. [CrossRef]
9. Pino, J.A.; Correa, M.T. Chemical Composition of the Essential Oil from Annatto (*Bixa orellana* L.) Seeds. *J. Essent. Oil Res.* **2003**, *15*, 66. [CrossRef]
10. Yong, Y.K.; Zakaria, Z.A.; Kadir, A.A.; Somchit, M.N.; Lian, G.E.C.; Ahmad, Z. Chemical constituents and antihistamine activity of *Bixa orellana* leaf extract. *BMC Complement. Altern. Med.* **2013**, *13*, 32. [CrossRef]
11. Parveen, M.; Malla, A.M.; Ali, A.; Nami, S.A.A.; Silva, P.S.P.; Silva, M.R. Isolation, Characterization, Bioassay and X-ray Crystallographic Study of Phytoconstituents from *Bixa orellana* Leaves. *Chem. Nat. Compd.* **2015**, *51*, 62. [CrossRef]
12. Khan, M.S.Y.; Sharma, S.; Husain, A. Synthesis and antibacterial evaluation of new flavonoid derivatives from 4,6-diacetyl resorcinol. *Sci. Pharm.* **2002**, *70*, 287. [CrossRef]
13. SADABS. *Area-Detector Absorption Correction*; Siemens Industrial Automation, Inc.: Madison, WI, USA, 1996.
14. Sheldrick, G.M. *SHELXL-97: Program for Crystal Structure Refinement*; University of Göttingen: Göttingen, Germany, 1997.
15. Platon, S.A. An Integrated Tool for the Analysis of the Results of a Single Crystal Structure Determination. *Acta Crystallogr. Sect. A* **1990**, *46*, C34.
16. Platon, S.A. *A Multipurpose Crystallographic Tool*; Utrecht University: Utrecht, The Netherlands, 1998.
17. Farrugia, L.J. WinGX and ORTEP for Windows: An update. *J. Appl. Cryst.* **2012**, *45*, 849. [CrossRef]
18. Grime, S.; Antony, J.; Ehrlich, S.; Krieg, H. A consistent and accurate ab initio parametrization of density functional dispersion correction (DFT-D) for the 94 elements H-Pu. *J. Chem. Phys.* **2010**, *132*, 154104. [CrossRef] [PubMed]
19. Frisch, M.J.; Trucks, G.W.; Schlegel, H.B.; Scuseria, G.E.; Robb, M.A.; Cheeseman, J.R.; Montgomery, J.A., Jr.; Vreven, T.; Kudin, K.N.; Burant, J.C.; et al. *Gaussian 03, Revision E.01*; Gaussian Inc.: Wallingford, CT, USA, 2004.
20. Schmidt, J.R.; Polik, W.F. *WebMO Enterprise, 18.1.001*; WebMO LLC: Holland, MI, USA, 2016.
21. Dennington, R.; Keith, T.; Millam, J. *Gauss View Version 5*; Semichem Inc.: Shawnee Mission, AR, USA, 2009.
22. Chemcraft-Graphical Software for Visualization of Quantum Chemistry Computations. Available online: https://www.chemcraftprog.com (accessed on 1 November 2021).
23. O'boyle, N.M.; Tenderholt, A.L. cclib: A library for package-independent computational chemistry algorithms. *J. Comput. Chem.* **2008**, *29*, 839. [CrossRef] [PubMed]
24. Jayaram, B.; Singh, T.; Mukherjee, G.; Mathur, A.; Shekhar, S.; Shekhar, V. Sanjeevini: A freely accessible webserver for target directed lead molecule discovery. *BMC Bioinform.* **2012**, *13*, 1. [CrossRef] [PubMed]
25. Daina, A.; Michielin, O.; Zoete, V. SwissADME: A free web tool to evaluate pharmacokinetics, drug-likeness and medicinal chemistry friendliness of small molecules. *Sci. Rep.* **2017**, *7*, 42717. [CrossRef]
26. Kokila, M.K.; Nirmala, K.A.; Shamala, P.N. Structure of 4,6-di acetylresorcinol. *Acta. Cryst.* **1992**, *C48*, 1133.
27. Rossi, M.; Erlebacher, J.; Zacharias, D.E.; Carrell, H.L.; Iannucci, B. The crystal and molecular structure of ellagic acid dihydrate: A dietary anti-cancer agent. *Carcinogenesis* **1991**, *12*, 2227. [CrossRef]
28. Socrates, G. *Infrared Characteristic Group Frequencies*, 3rd ed.; Wiley Interscience Publications: New York, NY, USA, 1980.
29. Kosar, B.; Albayrak, C. Spectroscopic investigations and quantum chemical computational study of (E)-4-methoxy-2-[(p-tolylimino)methyl]phenol. *Spectrochim. Acta A* **2011**, *78*, 160. [CrossRef]

30. Alam, M.J.; Ahmad, S. Quantum chemical and spectroscopic investigations of 3-methyladenine. *Spectrochim. Acta A* **2014**, *128*, 653. [CrossRef] [PubMed]
31. Tabatchnik, A.; Blot, V.; Pipelier, M.; Dubreuil, D.; Renault, E.; Le Questel, J.-Y. Theoretical Study of the Structures and Hydrogen-Bond Properties of New Alternated Heterocyclic Compounds. *J. Phys. Chem. A* **2010**, *114*, 6413. [CrossRef] [PubMed]
32. Afrin, S.; Rahman, Y.; Sarwar, T.; Husain, M.A.; Ali, A.; Tabish, M. Molecular spectroscopic and thermodynamic studies on the interaction of anti-platelet drug ticlopidine with calf thymus DNA. *Spectrochim. Acta Part A Mol. Biomol. Spectrosc.* **2017**, *186*, 66. [CrossRef] [PubMed]
33. Parveen, M.; Aslam, A.; Siddiqui, S.; Tabish, M.; Alam, M. Structure elucidation, DNA binding and molecular docking studies of natural compounds isolated from Crateva religiosa leaves. *J. Mol. Struct.* **2021**, *1251*, 131976. [CrossRef]
34. Sirajuddin, M.; Ali, S.; Badshah, A. Drug-DNA interactions and their study by UV-Visible, fluorescence spectroscopies and cyclic voltametry. *J. Photochem. Photobiol. B Biol.* **2013**, *124*, 1–19. [CrossRef]
35. Hussain, I.; Fatima, S.; Siddiqui, S.; Ahmed, S.; Tabish, M. Exploring the binding mechanism of β-resorcylic acid with calf thymus DNA: Insights from multi-spectroscopic, thermodynamic and bioinformatics approaches. *Spectrochim. Acta Part A Mol. Biomol. Spectrosc.* **2021**, *260*, 119952. [CrossRef]
36. Siddiqui, S.; Ameen, F.; Jahan, I.; Nayeem, S.M.; Tabish, M. A comprehensive spectroscopic and computational investigation on the binding of the anti-asthmatic drug triamcinolone with serum albumin. *New J. Chem.* **2019**, *43*, 4137. [CrossRef]
37. Siddiqui, S.; Mujeeb, A.; Ameen, F.; Ishqi, H.M.; Rehman, S.U.; Tabish, M. Investigating the mechanism of binding of nalidixic acid with deoxyribonucleic acid and serum albumin: A biophysical and molecular docking approaches. *J. Biomol. Struct. Dyn.* **2021**, *39*, 570. [CrossRef]
38. Siddiqui, S.; Ameen, F.; ur Rehman, S.; Sarwar, T.; Tabish, M. Studying the interaction of drug/ligand with serum albumin. *J. Mol. Liq.* **2021**, *336*, 116200. [CrossRef]
39. Ameen, F.; Siddiqui, S.; Jahan, I.; Nayeem, S.M.; ur Rehman, S.; Tabish, M. A detailed insight into the interaction of memantine with bovine serum albumin: A spectroscopic and computational approach. *Spectrochim. Acta Part A Mol. Biomol. Spectrosc.* **2022**, *265*, 120391. [CrossRef]
40. Enmozhi, S.K.; Raja, K.; Sebastine, I.; Joseph, J. Andrographolide as a potential inhibitor of SARS-CoV-2 main protease: An in silico approach. *J. Biomol. Struct. Dyn.* **2021**, *39*, 3092. [CrossRef] [PubMed]
41. Benet, L.Z.; Hosey, C.M.; Ursu, O.; Oprea, T.I. BDDCS, the Rule of 5 and drugability. *Adv. Drug Deliv. Rev.* **2016**, *101*, 89. [CrossRef] [PubMed]
42. Pathak, K.; Raghuvanshi, S. Oral bioavailability: Issues and solutions via nanoformulations. *Clin. Pharmacokinet.* **2015**, *54*, 325. [CrossRef] [PubMed]
43. Zothantluanga, J.H.; Gogoi, N.; Shakya, A.; Chetia, D.; Lalthanzara, H. Computational guided identification of potential leads from *Acacia pennata* (L.) Willd. as inhibitors for cellular entry and viral replication of SARS-CoV-2. *Future J. Pharm. Sci.* **2021**, *7*, 201. [CrossRef]

Article

Crystal Structures, Thermal and Luminescent Properties of Gadolinium(III) *Trans*-1,4-cyclohexanedicarboxylate Metal-Organic Frameworks

Pavel A. Demakov [1,*], Alena A. Vasileva [1,2], Vladimir A. Lazarenko [3], Alexey A. Ryadun [1] and Vladimir P. Fedin [1]

[1] Nikolaev Institute of Inorganic Chemistry, Siberian Branch of the Russian Academy of Sciences, 630090 Novosibirsk, Russia; a.vasileva2@g.nsu.ru (A.A.V.); ryadunalexey@mail.ru (A.A.R.); cluster@niic.nsc.ru (V.P.F.)
[2] Department of Natural Sciences, Novosibirsk State University, 630090 Novosibirsk, Russia
[3] National Research Centre "Kurchatov Institute", 123182 Moscow, Russia; vladimir.a.lazarenko@gmail.com
* Correspondence: demakov@niic.nsc.ru

Citation: Demakov, P.A.; Vasileva, A.A.; Lazarenko, V.A.; Ryadun, A.A.; Fedin, V.P. Crystal Structures, Thermal and Luminescent Properties of Gadolinium(III) *Trans*-1,4-cyclohexanedicarboxylate Metal-Organic Frameworks. *Crystals* **2021**, *11*, 1375. https://doi.org/10.3390/cryst11111375

Academic Editor: Alexander Pöthig

Received: 27 October 2021
Accepted: 10 November 2021
Published: 11 November 2021

Publisher's Note: MDPI stays neutral with regard to jurisdictional claims in published maps and institutional affiliations.

Copyright: © 2021 by the authors. Licensee MDPI, Basel, Switzerland. This article is an open access article distributed under the terms and conditions of the Creative Commons Attribution (CC BY) license (https://creativecommons.org/licenses/by/4.0/).

Abstract: Four new gadolinium(III) metal-organic frameworks containing 2,2′-bipyridyl (bpy) or 1,10-phenanthroline (phen) chelate ligands and *trans*-1,4-cyclohexanedicarboxylate (chdc^{2-}) were synthesized. Their crystal structures were determined by single-crystal X-ray diffraction analysis. All four coordination frameworks are based on the binuclear carboxylate building units. In the compounds [Gd$_2$(bpy)$_2$(chdc)$_3$]·H$_2$O (**1**) and [Gd$_2$(phen)$_2$(chdc)$_3$]·0.5DMF (**2**), the six-connected {Ln$_2$(L)$_2$(OOCR)$_6$} blocks form a 3D network with the primitive cubic (pcu) topology. In the compounds [Gd$_2$(NO$_3$)$_2$(phen)$_2$(chdc)$_2$]·2DMF (**3**) and [Gd$_2$Cl$_2$(phen)$_2$(chdc)$_2$]·0.3DMF·2.2dioxane (**4**), the four-connected {Ln$_2$(L)$_2$(X)$_2$OOCR)$_4$} units (where X = NO$_3^-$ for **3** or Cl$^-$ for **4**) form a 2D square-grid (sql) network. The solid-state luminescent properties were investigated for the synthesized frameworks. Bpy-containing compound **1** shows no luminescence, possibly due to the paramagnetic quenching by Gd^{3+} cation. In contrast, the phenathroline-containing MOFs **2–4** possess yellow emission under visible excitation (λ_{ex} = 460 nm) with the tuning of the characteristic wavelength by the coordination environment of the metal center.

Keywords: metal-organic frameworks; coordination polymers; rare earth elements; gadolinium; luminescence

1. Introduction

Metal-organic frameworks (MOFs) are an important class of coordination compounds extensively studied in recent years. Their porosity as well as a wide variability of metal centers and organic ligands unveil a route to design materials with highly tunable adsorption, catalytic, optical and other physico-chemical properties. In particular, lanthanide(III) MOFs deserve a great interest due to the unique f^n electron configuration of the metal center and subsequent applications in magnetic and luminescent materials [1–9].

Gadolinium(III), having a half-filled f^7 sublevel, takes a special place in the lanthanide row. This is a most relevant paramagnetic center in the development of contrast agents for magnetic-resonance tomography and visualization [10–14]. The most intensive electron transitions in Gd^{3+} occur in the ultraviolet region of 290–318 nm with the narrow-banded emission and have been applied in common lasers. Under a soft UV and visible excitation, Gd(III) is a non-emissive center and the luminescence of its coordination compounds is ligand-centered [15–23].

Trans-1,4-cyclohexanedicarboxylate is quite rare, but is still a commercially available example of an aliphatic ligand, which saturated backbone is known to have extremely low UV/vis absorbance and no luminescent activity. Several examples of the distribution of such unusual optical properties to the coordination frameworks using aliphatic ligands have been reported in the literature to date [24–27]. In this work, four new gadolinium(III) *trans*-1,4-cyclohexanedicarboxyalte metal-organic frameworks based on the binuclear carboxylate blocks with the additionally coordinated N-donor chelate ligands were synthesized and characterized. The determined crystallographic formulas of the compounds are [Gd$_2$(bpy)$_2$(chdc)$_3$]·H$_2$O (**1**, bpy = 2,2'-bipyridyl), [Gd$_2$(phen)$_2$(chdc)$_3$]·0.5DMF (**2**, phen = 1,10-phenanthroline), [Gd$_2$(NO$_3$)$_2$(phen)$_2$(chdc)$_2$]·2DMF (**3**) and [Gd$_2$Cl$_2$(phen)$_2$(chdc)$_2$]·0.3DMF·2.2dioxane (**4**). A new molecular complex [Gd(DMF)$_2$(phen)]Cl$_3$ (**5**) not containing *trans*-1,4-cyclohexanedicarboxylate bridge (see Figure A1 in Appendix A) was crystallized during this work. The successful synthesis of such a series with a non-photoactive bridging ligand and a non-emissive paramagnetic metal center makes it possible to investigate the impact of the metal ion coordination environment on the N-donor ligand-centered luminescence in the corresponding coordination networks.

2. Materials and Methods

2.1. Materials

Trans-1,4-cyclohexanedicarboxylic acid (H$_2$chdc, >97.0%), 2,2'-bipyridyl (bpy, >99.0%) and 1,10-phenanthroline monohydrate (phen·H$_2$O, >98.0%) were received from TCI. Gd(NO$_3$)$_3$·6H$_2$O (99.9% REO) was received from Dalchem. GdCl$_3$·6H$_2$O (high-purity grade) was received from Krystall. N,N-dimethylformamide (DMF, reagent grade) and dioxane were received from Vekton.

2.2. Instruments

IR spectra in KBr pellets were recorded in the range 4000–400 cm^{-1} on a Bruker Scimitar FTS 2000 spectrometer. Elemental analysis was conducted with a VarioMICROcube analyzer. Powder X-ray diffraction (PXRD) analysis was performed at room temperature on a Shimadzu XRD-7000 diffractometer (Cu-Kα radiation, λ = 1.54178 Å, or Co-Kα radiation, λ = 1.78897 Å). Thermogravimetric analysis was carried out using a Netzsch TG 209 F1 Iris instrument under Ar flow (30 cm^3·min^{-1}) at a 10 K·min^{-1} heating rate. Photoluminescence spectra were recorded with a spectrofluorometer Horiba Jobin Yvon Fluorolog 3 equipped with ozone-free Xe-lamp 450W power, cooled photon detector R928/1860 PFR technologies with refrigerated chamber PC177CE-010 and double grating monochromators. The spectra were corrected for source intensity and detector spectral response by standard correction curves. The absolute quantum yield was measured using a G8 (GMP SA, Switzerland) spectralon-coated integrating sphere, which was connected to a Fluorolog 3 spectrofluorometer. Diffraction data for single crystals of **1** were obtained on the 'Belok' beamline [28,29] (λ = 0.74539 Å) of the National Research Center 'Kurchatov Institute' (Moscow, Russian Federation) using a Rayonix SX165 CCD detector. The data were indexed, integrated and scaled, and absorption correction was applied using the XDS program package [30]. Diffraction data for single crystals of **2–5** were collected on an automated Agilent Xcalibur diffractometer equipped with an area AtlasS2 detector (graphite monochromator, λ(MoKα) = 0.71073 Å). Integration, absorption correction and determination of unit cell parameters were performed using the CrysAlisPro program package [31]. The structures were solved by the dual-space algorithm (SHELXT [32]) and refined by the full-matrix least squares technique (SHELXL [33]) in the anisotropic approximation (except hydrogen atoms). Positions of hydrogen atoms of organic ligands were calculated geometrically and refined in the riding model. The crystallographic data and details of the structure refinements are summarized in Appendix Table A1.

2.3. Synthetic Procedures

Synthesis of [Gd$_2$(bpy)$_2$(chdc)$_3$]·H$_2$O (**1**): 90.4 mg (0.20 mmol) of Gd(NO$_3$)$_3$·6H$_2$O, 62.4 mg (0.40 mmol) of bpy, 68.8 mg (0.40 mmol) of H$_2$chdc and 44.8 mg (0.80 mmol) of KOH were mixed in a 30 mL Teflon-lined stainless-steel autoclave and dispersed in 20 mL of H$_2$O. The obtained suspension was heated at 180 °C for 12 h. After cooling to room temperature, the obtained white precipitate was filtered off, washed with H$_2$O and dried in air. Yield: 79.2 mg (84%). IR spectrum main bands (KBr, cm^{-1}; see Figure S5): 3452 (s., br., νO–H); 3115 and 3088 (w., νCsp^2–H); 2927 and 2858 (m., νCsp^3–H); 1541 (s., νCOO$_{as}$); 1417 (s., νCOO$_s$). Elemental analysis data, calculated for [Gd$_2$(bpy)$_2$(chdc)$_3$]·(%): C, 46.5; H, 4.1; N, 4.9. Found (%): C, 46.4; H, 4.0; N, 4.8. PXRD data: Figure S1.

Synthesis of [Gd$_2$(phen)$_2$(chdc)$_3$]·0.5DMF (2): 148.8 mg (0.40 mmol) of GdCl$_3$·6H$_2$O, 72.0 mg (0.40 mmol) of phen·H$_2$O and 137 mg (0.80 mmol) of H$_2$chdc were mixed in a 20 mL glass vial and dissolved in 10 mL of DMF. The obtained solution was heated at 110 °C for 72 h. After cooling to room temperature, the obtained white precipitate was filtered off, washed with DMF and dried in air. Yield: 90.5 mg (37%). IR spectrum main bands (KBr, cm^{-1}): 3446 (w., br., νO–H); 3078 and 3062 (m., νCsp^2–H); 3010 and 2997 (w., νCsp^2–H); 2927 and 2856 (m., νCsp^3–H); 1685 (m., νCO$_{amide}$); 1598 and 1587 (s., νCOO$_{as}$); 1409 (s., νCOO$_s$). Elemental analysis data, calculated for [Gd$_2$(phen)$_2$(chdc)$_3$]·0.5DMF·H$_2$O (%): C, 47.9; H, 4.1; N, 5.1. Found (%): C, 47.9; H, 4.1; N, 5.4. TG data: 4% weight loss before 300 °C; calculated for 0.5DMF + H$_2$O: 4%. PXRD data: Figure S2.

Synthesis of [Gd$_2$(NO$_3$)$_2$(phen)$_2$(chdc)$_2$]·2DMF (3): 45.1 mg (0.10 mmol) of Gd(NO$_3$)$_3$·6H$_2$O, 18.0 mg (0.10 mmol) of phen·H$_2$O and 17.2 mg (0.10 mmol) of H$_2$chdc were mixed in a 5 mL glass vial and dissolved in 3.50 mL of DMF. Then, a 10 μL drop of HNO$_3$ (65%) was added. The obtained solution was heated at 110 °C for 36 h. After cooling to room temperature, the obtained white precipitate was filtered off, washed with DMF and dried in air. Yield: 43.7 mg (96%). IR spectrum main bands (KBr, cm^{-1}): 3420 (w., br., νO–H); 3078 and 3062 (m., νCsp^2–H); 2936 and 2858 (m., νCsp^3–H); 1681 (m., νCO$_{amide}$); 1587 (s., νCOO$_{as}$); 1426 (s., νCOO$_s$); 1292 (m., νNO). Elemental analysis data, calculated for [Gd$_2$(NO$_3$)$_2$(phen)$_2$(chdc)$_2$]·2DMF·2H$_2$O (%): C, 41.8; H, 4.1; N, 8.5. Found (%): C, 41.9; H, 4.0; N, 7.6. TG data: 13% weight loss at 120 °C; calculated for 2DMF + 2H$_2$O: 14%. PXRD data: Figure S3.

Synthesis of [Gd$_2$Cl$_2$(phen)$_2$(chdc)$_2$]·0.3DMF·2.2dioxane (4): 74.3 mg (0.20 mmol) of GdCl$_3$·6H$_2$O, 36.0 mg (0.20 mmol) of phen·H$_2$O and 34.4 mg (0.20 mmol) of H$_2$chdc were mixed in a 3 mL glass vial and dissolved in the mixture of 1 mL of DMF and 1.50 mL of 1,4-dioxane. Then, 25 μL of HCl (36%) was added. The obtained solution was heated at 100 °C for 36 h. After cooling to room temperature, the obtained white precipitate was filtered off, washed with DMF, then with dioxane, and dried in air. Yield: 54.0 mg (43%). IR spectrum main bands (KBr, cm^{-1}): 3392 (m., br., νO–H); 3080 and 3049 (w., νCsp^2–H); 2935 and 2856 (m., νCsp^3–H); 1654 (w., νCO$_{amide}$); 1598 and 1587 (s., νCOO$_{as}$); 1425 (s., νCOO$_s$). Elemental analysis data, calculated for [Gd$_2$Cl$_2$(phen)$_2$(chdc)$_2$]·0.3DMF·3H$_2$O (%): C, 42.3; H, 3.8; N, 5.2. Found (%): C, 41.8; H, 3.5; N, 5.4. TG data: 6% weight loss at 130 °C; calculated for 0.3DMF + 3H$_2$O: 6%. PXRD data: Figure S4.

3. Results and Discussion

3.1. Synthesis and Crystal Structure Description

Compounds [Gd$_2$(bpy)$_2$(chdc)$_3$]·H$_2$O (1), [Gd$_2$(phen)$_2$(chdc)$_3$]·0.5DMF (2) and [Gd$_2$(NO$_3$)$_2$(phen)$_2$(chdc)$_2$]·2DMF (3) crystallize in the monoclinic crystal system with $P2_1/n$ space group and are isostructural to the series previously reported by our group: [Ln$_2$(bpy)$_2$(chdc)$_3$]·xH$_2$O (x = 0 ... 1), [Ln$_2$(phen)$_2$(chdc)$_3$]·0.5DMF [34] and [Ln$_2$(NO$_3$)$_2$(phen)$_2$(chdc)$_2$]·2DMF [35] (Ln^{3+} = Y^{3+}, Eu^{3+} or Tb^{3+}), respectively. Highly effective luminescence with quantum yields up to 63% was revealed for these structures, including [Y$_2$(bpy)$_2$(chdc)$_3$] based on the non-emissive and diamagnetic Y^{3+}. Unlike the yttrium(III), Gd^{3+} is paramagnetic but still non-emissive in the soft UV and visible region; therefore, the investigation of isostructural Gd(III)-based compounds 1–3 and the cognate new metal-organic framework 4 could provide significantly new information concerning the luminescent properties of such type series of coordination networks.

Compounds 2 and 3 were obtained in quite similar solvothermal conditions at 110 °C, using N,N-dimethylformamide (DMF) as a solvent. The main difference between their synthetic methods is the starting Gd(III) salt (chloride for 2 or nitrate for 3) to be used. Thus, it was shown that the anion plays a main structure-forming role in this system, as the presence of NO$_3^-$ leads to the crystallization of nitrate-containing layered MOF 3. On the contrary, poorly coordinated Cl⁻ cannot compete with the carboxylate in such conditions and affords a three-dimensional framework **2**, which contains the only RCOO⁻ fragments additionally coordinated to the {Gd(phen)}$^{3+}$ moiety. Interestingly, the replacement of pure DMF by the mixture of DMF and dioxane (2:3) leads to the formation of the chloride-containing structure [Gd$_2$Cl$_2$(phen)$_2$(chdc)$_2$]·0.3DMF·2.2dioxane (**4**) instead of the carboxylate **2**. A significant dilution of DMF by dioxane apparently weakens the solvation of Cl⁻ and, therefore, strengthens its coordination ability compared to the carboxylate. Thus, the crystallization of two-dimensional framework **4** seems to be anion/solvent-controlled again and suggests a simple route for the synthesis of lanthanide(III) networks with coordinated halogenides, since only one example of the compound containing {Ln$_2$(Cl)$_2$(L)$_2$(OOCR)$_4$} building units (L = any 2,2′-bipyridyl derivative) has been reported to date [36].

In all the structures **1–3**, Gd(III) adopts a similar capped square-antiprismatic environment consisting of two N atoms of diimine chelate ligand and seven O atoms, which belong to the carboxylic groups in **1** and **2** or to carboxylic groups and terminal nitrate in **3** (Figure 1a–c). The selected coordination bond lengths are listed in Table 1. Two symmetry-equivalent Gd(III) ions form binuclear carboxylate blocks {Gd$_2$(bpy)$_2$(RCOO-κ^2)$_2$(µ-RCOO-κ^1,κ^1)$_2$(µ-RCOO-κ^1,κ^2)$_2$} (**1**), {Gd$_2$(phen)$_2$(RCOO-κ^2)$_2$(µ-RCOO-κ^1,κ^1)$_2$(µ-RCOO-κ^1,κ^2)$_2$} (**2**) and {Gd$_2$(phen)$_2$(ONO$_2$-κ^2)$_2$(µ-RCOO-κ^1,κ^1)$_2$(µ-RCOO-κ^1,κ^2)$_2$} (**3**). The coordination frameworks in **1** and **2**, consisting of six-connected building units and *trans*-1,4-cyclohexanedicarboxylate bridges, adopt a very distorted three-dimensional primitive cubic topology (pcu) and contain small voids (Figure 2a,b) interconnected by very narrow (2 × 2 Å^2) windows with 5% calculated total void volume. In the coordination framework in **3**, four-connected binuclear blocks are interconnected by cyclohexane moieties into a two-dimensional square-grid network (sql) (Figure 2c). The layers in **3** are packed in a one-layer (AA) manner to form channels with 26% general void volume. These channels are occupied by the localized DMF solvent molecules.

Table 1. Selected bond lengths in the structures **1–4**.

Bond	1	2	3	4
Gd–N, Å	2.566(2), 2.600(3)	2.499(19)–2.59(3)	2.553(2), 2.591(3)	2.564(7), 2.579(5)
Gd–O(COO$_{chelate}$), Å	2.3482(19)–2.514(2)	2.334(3)–2.539(3)	2.330(2)–2.573(2)	2.294(5)–2.513(5)
Gd–O(COO$_{non-chelate}$), Å or Gd–O(NO$_{3,non-chelate}$), Å	2.429(2)–2.4800(19)	2.463(3)–2.465(3)	2.458(2)–2.529(2)	-

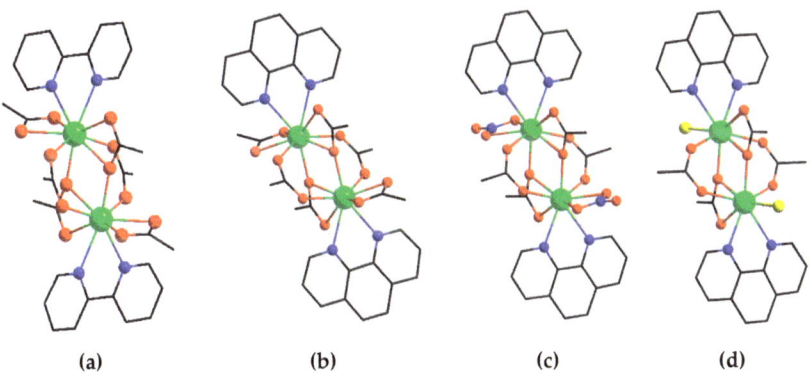

Figure 1. Secondary building units in **1** (**a**), **2** (**b**), **3** (**c**) and **4** (**d**). Gd atoms are green, N atoms are blue, O atoms are red, Cl atoms are yellow. H atoms are not shown.

Compound [Gd$_2$Cl$_2$(phen)$_2$(chdc)$_2$]·0.3DMF·2.2dioxane (**4**) crystallizes in the monoclinic crystal system with the $P2_1/n$ space group. The coordination environment of Gd(III) consists of two N atoms of diimine chelate ligand, five O atoms of the carboxylic groups and one Cl atom. The Gd–N and Gd–O bond lengths are close to those in **1–3** (see Table 1) and the Gd–Cl distance is 2.649(2) Å (Figures 1 and 2). The structure of a binuclear carboxylate block {Gd$_2$(phen)$_2$(Cl)$_2$(μ-RCOO-κ^1,κ^1)$_2$(μ-RCOO-κ^1,κ^2)$_2$} in **4** is analogous to the nitrate-containing unit in **3**, except for the reduction of the coordination number to 8 due to the substitution of the bidentate nitrate anion with the larger chloride (Figure 1d), which acts as a monodentate ligand. Four-connected binuclear blocks in **4** are interconnected by cyclohexane moieties (Figure 2d) in a similar AA manner to **3**, with the channels of 32% general void volume. These channels are occupied by solvent molecules. Only one dioxane molecule per formula unit was localized directly, while the non-ordered residual electron density was analyzed by the PLATON/SQUEEZE [37] procedure (69 e$^-$ in 265 Å per formula unit) and assigned to 0.3DMF + 1.2dioxane (69.6 e$^-$ and ca. 209 Å volume estimated from the liquid densities).

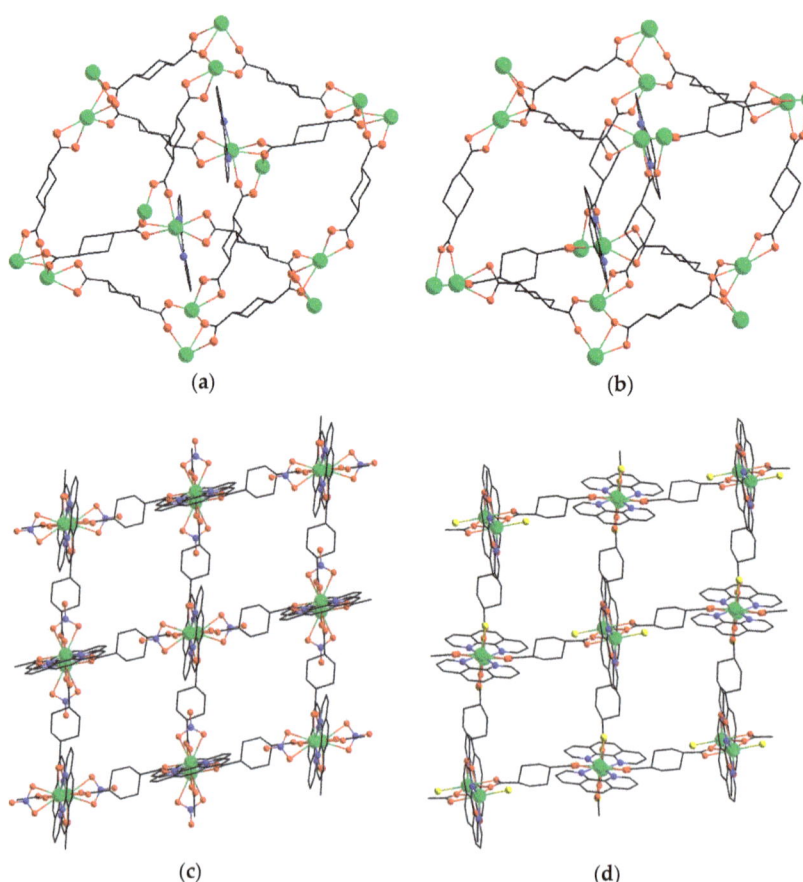

Figure 2. Fragments of three-dimensional coordination frameworks in **1** (**a**) and **2** (**b**). Coordination layers in **3** (**c**) and **4** (**d**), Atom colors are similar to those in Figure 1.

3.2. Thermal Properties

Thermogravimetric analyses for the compounds **1–4** were performed (Figure 3). For **1**, the stepwise decomposition starts at ca. 350 °C. Only 1% weight loss before 300 °C corresponds well to the low content of guest solvent molecules (calculated as 1.5%) determined by X-ray crystallography, and 33% residual weight at 600 °C matches well to the gadolinium(III) oxide (calculated as 32%).

2 slowly losses solvent molecules at the temperature up to 300 °C, much higher than the boiling points of both DMF and water. Such feature is apparently attributed to the low size of the windows (~2 Å) in the coordination framework of **2** and the resulting kinetic hindrance of the guest diffusion. The first step of lattice decomposition occurs in the range 340–440 °C and corresponds well to the loss of phen molecules (66% residue at 440 °C; calculated for Gd$_2$(chdc)$_3$: 66.5%). Further weight loss starting at ca. 460 °C corresponds to the decomposition of the bridging ligand and leads to the Gd$_2$O$_3$ (34% residue at 600 °C, calculated: 29%) being apparently contaminated by carbon admixture due to the incomplete evaporation of the organic moieties.

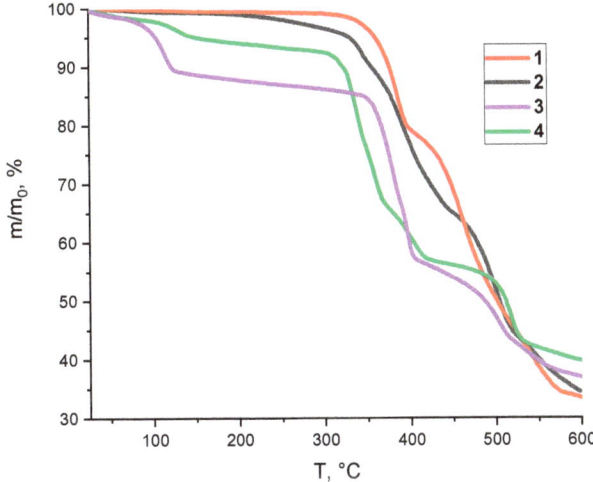

Figure 3. TG plots for compounds **1–4**.

Compound **3** loses solvents at ca. 120 °C. The first step of coordination lattice decomposition occurs in the range 350–400 °C and corresponds well to the loss of phen molecules (58% residue at 440 °C; calculated for Gd$_2$(NO$_3$)$_2$(chdc)$_2$: 59%). Further weight loss, corresponding to the decomposition of the nitrate and chdc^{2-} ligands, starts at ca. 430 °C. The TG profile of **4** is close to **3** and includes solvent loss at ca. 130 °C and two steps of the lattice decomposition occurring in the range 320–410 °C and above 490 °C. In summary, thermal stability characteristics of the coordination framework in **1–4** are high [38–41] and quite similar to each other. However, their stability is limited by the evaporation of the neutral N-donor chelate, occurring below 400 °C.

3.3. Luminescence Spectrocopy

Solid-state luminescence measurements were performed for the synthesized compounds. The bpy-containing **1** appeared to possess no luminescent activity, possibly due to the paramagnetic quenching of the emission by the Gd(III) cation. In contrast, the phen-containing compounds **2–4** demonstrate yellow wide-banded emission under a visible light excitation at λ_{ex} = 460 nm. The corresponding emission spectra are shown in Figure 4a. The maxima of the spectra appear at λ = 537 nm for three-dimensional **2** based on the six-connected carboxylate units, λ = 522 nm for the nitrate-containing **3** and λ = 556 nm for the chloride-containing **4**. The observed red-shift of both the maxima and the characteristic wavelengths in the row nitrate < carboxylate < chloride apparently correlates to the electron donor properties of the corresponding ligands. The calculated (x,y) coordinates on the CIE 1931 chromaticity diagram and characteristic wavelength values are shown in Figure 4b and visualize the integral yellow color of the wide-banded emission of **2–4**.

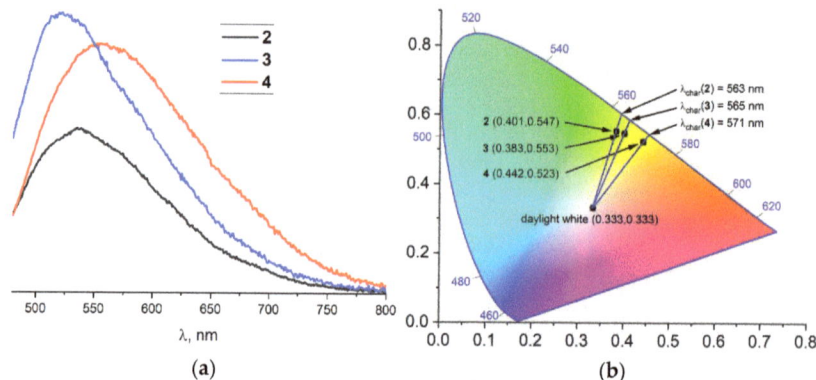

Figure 4. Emission spectra for **2–4** at λ_{ex} = 460 nm (**a**), CIE 1931 chromaticity diagram for **2–4** (**b**).

4. Conclusions

To summarize, four new gadolinium(III) metal-organic frameworks were synthesized and characterized. Compounds **1** and **2** containing six-connected binuclear metal-carboxylate blocks adopt a distorted primitive cubic topology with narrow pores. A partial substitution of carboxylate by nitrate or chloride in the Gd(III) coordination sphere leads to two-dimensional square-layered networks **3** and **4**. Thermal and luminescent properties of the synthesized compounds were investigated. Phenanthroline-based structures **2–4** emit in the yellow region under a visible blue excitation at 460 nm. The observed red-shift in the row of coordinated ligands, nitrate < carboxylate < chloride was attributed to the donor ability of the corresponding ligands.

Supplementary Materials: The following are available online at https://www.mdpi.com/article/10.3390/cryst11111375/s1: Figure S1: Experimental PXRD pattern for **1** compared to the theoretical one; Figure S2: Experimental PXRD pattern for **2** compared to the theoretical one; Figure S3: Experimental PXRD pattern for **3** compared to the theoretical one; Figure S4: Experimental PXRD pattern for **4** compared to the theoretical one; Figure S5: IR spectra for **1–5**.

Author Contributions: P.A.D., original draft preparation, single-crystal XRD, graphing; A.A.V., synthesis, characterization, graphing; V.A.L., synchrotron single-crystal XRD; A.A.R., solid-state luminescence measurements; V.P.F., manuscript review and editing, project administration and funding acquisition. All authors have read and agreed to the published version of the manuscript.

Funding: The research was supported by the Ministry of Science and Higher Education of the Russian Federation, No. 121031700321-3 and No. 121031700321-8.

Institutional Review Board Statement: Not applicable.

Informed Consent Statement: Not applicable.

Data Availability Statement: CCDC 2118161–2118165 contain the supplementary crystallographic data for this paper. These data can be obtained free of charge from The Cambridge Crystallographic Data Center at https://www.ccdc.cam.ac.uk/structures/.

Conflicts of Interest: The authors declare no conflict of interest.

Appendix A. Crystal Structure of a New Complex [Gd(DMF)2(phen)]Cl3 (5)

Single crystals of **5** were obtained during the screening syntheses: 37.2 mg (0.10 mmol) of $GdCl_3 \cdot 6H_2O$, 18.0 mg (0.10 mmol) of phen·H_2O and 34.4 mg (0.20 mmol) of H_2chdc were mixed in a 3 mL glass vial and dissolved in the mixture of 1 mL of DMF and 1.50 mL of acetone. The obtained solution was heated at 45 °C for 72 h. After cooling to room temperature, the obtained white precipitate was filtered off, washed with DMF and dried in air. IR spectrum main bands (KBr, cm^{-1}): 3375 (s., br., νO−H); 2937 and 2858 (w., νCsp^3−H); 1654 (s., νCO_{amide}).

A molecular complex [Gd(DMF)$_2$(phen)]Cl$_3$ (**5**) crystallizes in the triclinic crystal system with the P-1 space group with one independent Gd atom and Z = 2. The coordination environment of Gd(III) consists of two N atoms of phenanthroline, two O atoms of the coordinated solvent and three Cl atoms. The Gd—N bond length is 2.584(2) Å, the Gd—O bond lengths are 2.334(2) Å and 2.374(2) Å and the Gd—Cl bond lengths are 2.6408(8) Å, 2.6694(8) Å and 2.6921(7) Å, which are close to the distance in the structure of **4**: 2.649(2) Å.

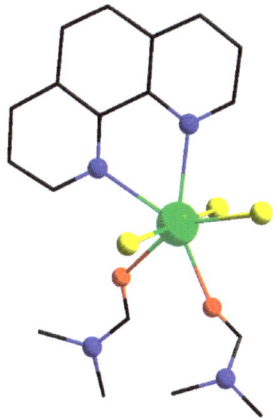

Figure A1. A molecular complex [Gd(DMF)$_2$(phen)]Cl$_3$. Gd atoms are green, N atoms are blue, O atoms are red, Cl atoms are yellow. H atoms are not shown.

Appendix B. The Crystallographic Data for 1–5

Table A1. Crystallographic data and refinement details for the structures 1–5.

	1	2	3	4	5
Chemical formula	C$_{44}$H$_{48}$Gd$_2$N$_4$O$_{13}$	C$_{49.5}$H$_{49.5}$Gd$_2$N$_{4.50}$O$_{12.5}$	C$_{46}$H$_{50}$Gd$_2$N$_8$O$_{16}$	C$_{47.7}$H$_{51.7}$Cl$_2$Gd$_2$N$_{4.3}$O$_{11.7}$	C$_{18}$H$_{22}$Cl$_3$GdN$_4$O$_2$
M$_r$, g/mol	1155.36	1221.93	1285.44	1257.83	589.99
Crystal system	Monoclinic	Monoclinic	Monoclinic	Monoclinic	Triclinic
Space group	$P2_1/n$	$P2_1/n$	$P2_1/n$	$P2_1/n$	$P\overline{1}$
Temperature, K	100	140	140	140	140
a, Å	10.422(2)	10.4479(4)	12.3986(6)	13.2219(5)	9.2772(4)
b, Å	17.551(4)	17.9554(7)	16.8287(6)	13.3512(7)	9.6641(4)
c, Å	12.294(3)	12.8866(7)	12.7182(5)	15.5831(7)	12.2954(5)
α, °	90	90	90	90	102.460(4)
b, °	103.90(3)	99.104(4)	112.863(5)	109.100(5)	94.032(3)
γ, °	90	90	90	90	92.587(3)
V, Å^3	2183.0(8)	2387.02(19)	2445.20(19)	2599.4(2)	1071.67(8)
Z	2	2	2	2	2
F(000)	1144	1212	1276	1247	578
D(calc.), g/cm^3	1.758	1.700	1.746	1.607	1.828
m, mm^{-1}	3.462	2.82	2.77	2.69	3.49
Crystal size, mm	0.20 × 0.04 × 0.04	0.30 × 0.14 × 0.10	0.30 × 0.21 × 0.11	0.25 × 0.17 × 0.14	0.65 × 0.25 × 0.20
θ range for data collection, °	2.42 ≤ θ ≤ 26.68	2.3 ≤ θ ≤ 25.4	2.0 ≤ θ ≤ 25.4	2.1 ≤ θ ≤ 25.4	2.2 ≤ θ ≤ 25.3

Table A1. *Cont.*

	1	2	3	4	5
θ range for data collection, °	$2.42 \leq \theta \leq 26.68$	$2.3 \leq \theta \leq 25.4$	$2.0 \leq \theta \leq 25.4$	$2.1 \leq \theta \leq 25.4$	$2.2 \leq \theta \leq 25.3$
No. of reflections: measured/independent/observed [$I > 2\sigma(I)$]	14,396/3987/3615	10,535/4383/3520	11,222/4487/3921	11,529/4755/3827	9676/3919/3637
R_{int}	0.0356	0.0249	0.0251	0.0231	0.0368
Index ranges	$-12 \leq h \leq 12$ $-21 \leq k \leq 20$ $-14 \leq l \leq 14$	$-12 \leq h \leq 12$ $-21 \leq k \leq 19$ $-15 \leq l \leq 15$	$-14 \leq h \leq 12$ $-15 \leq k \leq 20$ $-12 \leq l \leq 15$	$-11 \leq h \leq 15$ $-16 \leq k \leq 14$ $-18 \leq l \leq 18$	$-11 \leq h \leq 9$ $-11 \leq k \leq 11$ $-14 \leq l \leq 14$
Final R indices [$I > 2\sigma(I)$]	$R_1 = 0.0205$ $wR_2 = 0.0482$	$R_1 = 0.0286$ $wR_2 = 0.0630$	$R_1 = 0.0227$ $wR_2 = 0.0480$	$R_1 = 0.0475$ $wR_2 = 0.1365$	$R_1 = 0.0231$ $wR_2 = 0.0459$
R indices (all data)	$R_1 = 0.0240$ $wR_2 = 0.0496$	$R_1 = 0.0437$ $wR_2 = 0.0673$	$R_1 = 0.0296$ $wR_2 = 0.0502$	$R_1 = 0.0572$ $wR_2 = 0.1443$	$R_1 = 0.0261$ $wR_2 = 0.0470$
Goodness-of-fit on F^2	1.026	1.043	1.061	1.070	1.049
Largest diff. peak/hole, $e/Å^3$	0.525, −0.601	0.73, −0.80	0.66, −0.78	2.36, −0.86	0.63, −0.68

References

1. Nonat, A.M.; Charbonnière, L.J. Upconversion of light with molecular and supramolecular lanthanide complexes. *Coord. Chem. Rev.* **2020**, *409*, 213192. [CrossRef]
2. Belousov, Y.A.; Drozdov, A.A.; Taydakov, I.V.; Marchetti, F.; Pettinari, R.; Pettinari, C. Lanthanide azolecarboxylate com-pounds: Structure, luminescent properties and applications. *Coord. Chem. Rev.* **2021**, *445*, 214084. [CrossRef]
3. Xu, H.; Cao, C.-S.; Kang, X.-M.; Zhao, B. Lanthanide-based metal–organic frameworks as luminescent probes. *Dalton Trans.* **2016**, *45*, 18003–18017. [CrossRef] [PubMed]
4. Zhao, S.-N.; Wang, G.; Poelman, D.; Van Der Voort, P. Luminescent Lanthanide MOFs: A Unique Platform for Chemical Sensing. *Materials* **2018**, *11*, 572. [CrossRef] [PubMed]
5. Hao, Y.; Chen, S.; Zhou, Y.; Zhang, Y.; Xu, M. Recent Progress in Metal–Organic Framework (MOF) Based Luminescent Che-modosimeters. *Nanomaterials* **2019**, *9*, 974. [CrossRef] [PubMed]
6. Zhan, Z.; Jia, Y.; Li, D.; Zhang, X.; Hu, M. A water-stable terbium-MOF sensor for the selective, sensitive, and recyclable detection of Al^{3+} and CO_3^{2-} ions. *Dalton Trans.* **2019**, *48*, 15255–15262. [CrossRef]
7. Saraci, F.; Quezada-Novoa, V.; Donnarumma, P.R.; Howarth, A.J. Rare-earth metal–organic frameworks: From structure to applications. *Chem. Soc. Rev.* **2020**, *49*, 7949–7977. [CrossRef] [PubMed]
8. Litvinova, Y.M.; Gayfulin, Y.M.; Samsonenko, D.G.; Dorovatovskiy, P.V.; Lazarenko, V.A.; Brylev, K.A.; Mironov, Y.V. Coordination polymers based on rhenium octahedral chalcocyanide cluster [Re6Se8(CN)6]4− and lanthanide ions solvated with dimethylformamide. *Inorg. Chim. Acta* **2021**, *528*, 120597. [CrossRef]
9. Yao, C.-X.; Zhao, N.; Liu, J.-C.; Chen, L.-J.; Liu, J.-M.; Fang, G.-Z.; Wang, S. Recent Progress on Luminescent Metal-Organic Framework-Involved Hybrid Materials for Rapid Determination of Contaminants in Environment and Food. *Polymers* **2020**, *12*, 691. [CrossRef] [PubMed]
10. Caravan, P. Protein-Targeted Gadolinium-Based Magnetic Resonance Imaging (MRI) Contrast Agents: Design and Mechanism of Action. *Accounts Chem. Res.* **2009**, *42*, 851–862. [CrossRef] [PubMed]
11. Narmani, A.; Farhood, B.; Haghi-Aminjan, H.; Mortezazadeh, T.; Aliasgharzadeh, A.; Mohseni, M.; Najafi, M.; Abbasi, H. Gadolinium nanoparticles as diagnostic and therapeutic agents: Their delivery systems in magnetic resonance imaging and neutron capture therapy. *J. Drug Deliv. Sci. Technol.* **2018**, *44*, 457–466. [CrossRef]
12. Clough, T.J.; Jiang, L.; Wong, K.-L.; Long, N.J. Ligand design strategies to increase stability of gadolinium-based magnetic resonance imaging contrast agents. *Nat. Commun.* **2019**, *10*, 1420. [CrossRef] [PubMed]
13. Zairov, R.; Pizzanelli, S.; Dovzhenko, A.P.; Nizameev, I.; Orekhov, A.S.; Arkharova, N.; Podyachev, S.N.; Sudakova, S.; Mustafina, A.R.; Calucci, L. Paramagnetic Relaxation Enhancement in Hydrophilic Colloids Based on Gd(III) Complexes with Tetrathia- and Calix[4]arenes. *J. Phys. Chem. C* **2020**, *124*, 4320–4329. [CrossRef]
14. Parker, D.; Suturina, E.; Kuprov, I.; Chilton, N.F. How the Ligand Field in Lanthanide Coordination Complexes Determines Magnetic Susceptibility Anisotropy, Paramagnetic NMR Shift, and Relaxation Behavior. *Accounts Chem. Res.* **2020**, *53*, 1520–1534. [CrossRef]
15. Tan, Q.-H.; Wang, Y.-Q.; Guo, X.-Y.; Liu, H.-T.; Liu, Z.-L. A gadolinium MOF acting as a multi-responsive and highly selective luminescent sensor for detecting o-, m-, and p-nitrophenol and Fe^{3+} ions in the aqueous phase. *RSC Adv.* **2016**, *6*, 61725–61731. [CrossRef]

16. Casanovas, B.; Speed, S.; Maury, O.; El Fallah, M.S.; Font-Bardí, M.; Vicente, R. Dinuclear LnIII Complexes with 9-Anthracenecarboxylate Showing Field-Induced SMM and Visible/NIR Luminescence. *Eur. J. Inorg. Chem.* **2018**, *34*, 3859–3867. [CrossRef]
17. Casanovas, B.; Speed, S.; Maury, O.; Font-Bardia, M.; Vicente, R. Homodinuclear lanthanide 9-anthracenecarboxylate complexes: Field induced SMM and NIR-luminescence. *Polyhedron* **2019**, *169*, 187–194. [CrossRef]
18. Casanovas, B.; Speed, S.; Vicente, R.; Font-Bardia, M. Sensitization of visible and NIR emitting lanthanide(III) ions in a series of dinuclear complexes of formula [Ln$_2$(μ-2-FBz)$_2$(2-FBz)$_4$(terpy)$_2$]·2(2-HFBz)·2(H$_2$O). *Polyhedron* **2019**, *173*, 114113. [CrossRef]
19. Taydakov, I.V.; Belousov, Y.A.; Lyssenko, K.A.; Varaksina, E.; Drozdov, A.A.; Marchetti, F.; Pettinari, R.; Pettinari, C. Synthesis, phosphorescence and luminescence properties of novel europium and gadolinium tris-acylpyrazolonate complexes. *Inorg. Chim. Acta* **2019**, *502*, 119279. [CrossRef]
20. Bryleva, Y.A.; Artem'Ev, A.V.; Glinskaya, L.A.; Komarov, V.Y.; Bogomyakov, A.S.; Rakhmanova, M.I.; Larionov, S.V. A series of bis(2-phenethyl)dithiophosphinate-based Ln(III) complexes: Synthesis, magnetic and photoluminescent properties. *Inorg. Chim. Acta* **2021**, *516*, 120097. [CrossRef]
21. Kim, J.H.; Lepnev, L.S.; Utochnikova, V.V. Dual vis-NIR emissive bimetallic naphthoates of Eu–Yb–Gd: A new approach toward Yb luminescence intensity increase through Eu → Yb energy transfer. *Phys. Chem. Chem. Phys.* **2021**, *23*, 7213–7219. [CrossRef]
22. Gontcharenko, V.; Kiskin, M.; Dolzhenko, V.; Korshunov, V.; Taydakov, I.; Belousov, Y. Mono- and Mixed Metal Complexes of Eu^{3+}, Gd^{3+}, and Tb^{3+} with a Diketone, Bearing Pyrazole Moiety and CHF$_2$-Group: Structure, Color Tuning, and Kinetics of Energy Transfer between Lanthanide Ions. *Molecules* **2021**, *26*, 2655. [CrossRef] [PubMed]
23. Utochnikova, V.V.; Aslandukov, A.N.; Vashchenko, A.A.; Goloveshkin, A.S.; Alexandrov, A.A.; Grzibovskis, R.; Bunzli, J.-C.G. Identifying lifetime as one of the key parameters responsible for the low brightness of lanthanide-based OLEDs. *Dalton Trans.* **2021**, *50*, 12806–12813. [CrossRef] [PubMed]
24. Llabres-Campaner, P.J.; Pitarch-Jarque, J.; Ballesteros-Garrido, R.; Abarca, B.; Ballesteros, R.; García-España, E. Bicyclo[2.2.2]octane-1,4-dicarboxylic acid: Towards transparent metal-organic frameworks. *Dalton Trans.* **2017**, *46*, 7397–7402. [CrossRef]
25. Demakov, P.A.; Sapchenko, S.A.; Samsonenko, D.G.; Dybtsev, D.N.; Fedin, V.P. Coordination polymers based on zinc(ii) and manganese(ii) with 1,4-cyclohexanedicarboxylic acid. *Russ. Chem. Bull.* **2018**, *67*, 490–496. [CrossRef]
26. Yin, J.; Yang, H.; Fei, H. Robust, Cationic Lead Halide Layered Materials with Efficient Broadband White-Light Emission. *Chem. Mater.* **2019**, *31*, 3909–3916. [CrossRef]
27. Demakov, P.A.; Poryvaev, A.S.; Kovalenko, K.A.; Samsonenko, D.G.; Fedin, M.V.; Fedin, V.P.; Dybtsev, D.N. Structural Dy-namics and Adsorption Properties of the Breathing Microporous Aliphatic Metal–Organic Framework. *Inorg. Chem.* **2020**, *59*, 15724–15732. [CrossRef]
28. Svetogorov, R.; Dorovatovskii, P.V.; Lazarenko, V.A. Belok/XSA Diffraction Beamline for Studying Crystalline Samples at Kurchatov Synchrotron Radiation Source. *Cryst. Res. Technol.* **2020**, *55*, 1900184. [CrossRef]
29. Lazarenko, V.A.; Dorovatovskii, P.V.; Zubavichus, Y.V.; Burlov, A.S.; Koshchienko, Y.V.; Vlasenko, V.G.; Khrustalev, V.N. High-Throughput Small-Molecule Crystallography at the 'Belok' Beamline of the Kurchatov Synchrotron Radiation Source: Transition Metal Complexes with Azomethine Ligands as a Case Study. *Crystals* **2017**, *7*, 325. [CrossRef]
30. Kabsch, W. XDS. *Acta Crystallogr.* **2010**, *D66*, 125–132. [CrossRef]
31. CrysAlisPro, Version: 1.171.38.46. Rigaku Oxford Diffraction; Rigaku Americas Holding Company, Inc.: The Woodlands, TX, USA, 2015.
32. Sheldrick, G.M. SHELXT—Integrated space-group and crystal-structure determination. *Acta Crystallogr. Sect. A Found. Adv.* **2015**, *71*, 3–8. [CrossRef]
33. Sheldrick, G.M. Crystal structure refinement with SHELXL. *Acta Crystallogr. Sect. C Struct. Chem.* **2015**, *71*, 3–8. [CrossRef] [PubMed]
34. Demakov, P.A.; Ryadun, A.A.; Dorovatovskii, P.V.; Lazarenko, V.A.; Samsonenko, D.G.; Brylev, K.A.; Fedin, V.P.; Dybtsev, D.N. Intense multi-colored luminescence in a series of rare-earth metal–organic frameworks with aliphatic linkers. *Dalton Trans.* **2021**, *50*, 11899–11908. [CrossRef] [PubMed]
35. Demakov, P.A.; Vasileva, A.A.; Volynkin, S.S.; Ryadun, A.A.; Samsonenko, D.G.; Fedin, V.P.; Dybtsev, D.N. Cinnamal Sensing and Luminescence Color Tuning in a Series of Rare-Earth Metal–Organic Frameworks with Trans-1,4-cyclohexanedicarboxylate. *Molecules* **2021**, *26*, 5145. [CrossRef]
36. Lu, Y.-B.; Jiang, X.-M.; Zhu, S.-D.; Du, Z.-Y.; Liu, C.-M.; Xie, Y.-R.; Liu, L.-X. Anion Effects on Lanthanide(III) Tetrazole-1-acetate Dinuclear Complexes Showing Slow Magnetic Relaxation and Photofluorescent Emission. *Inorg. Chem.* **2016**, *55*, 3738–3749. [CrossRef] [PubMed]
37. Spek, A.L. Single-crystal structure validation with the program PLATON. *J. Appl. Crystallogr.* **2003**, *36*, 7–13. [CrossRef]
38. Barsukova, M.; Samsonenko, D.G.; Fedin, V.P. Crystal structure of metal-organic frameworks based on terbium and 1,4-naphthalenedicarboxylic acid. *J. Struct. Chem.* **2020**, *61*, 1090–1096. [CrossRef]
39. Demakov, P.A.; Ryadun, A.A.; Samsonenko, D.G.; Dybtsev, D.N.; Fedin, V.P. Structure and luminescent properties of europium(III) coordination polymers with thiophene ligands. *J. Struct. Chem.* **2020**, *61*, 1965–1974. [CrossRef]

40. Barsukova, M.O.; Cherezova, S.V.; Sapianik, A.A.; Lundovskaya, O.V.; Samsonenko, D.G.; Fedin, V.P. Lanthanide contraction effect and white-emitting luminescence in a series of metal–organic frameworks based on 2,5-pyrazinedicarboxylic acid. *RSC Adv.* **2020**, *10*, 38252–38259. [CrossRef]
41. Cherezova, S.V.; Barsukova, M.O.; Samsonenko, D.G.; Fedin, V.P. Crystal structure of dense metal-organic frameworks based on sc(III) and two types of ligands. *J. Struct. Chem.* **2021**, *62*, 897–904. [CrossRef]

Article

Influence of Organic-Modified Inorganic Matrices on the Optical Properties of Palygorskite–Curcumin-Type Hybrid Materials

Florentina Monica Raduly, Valentin Răditoiu *, Radu Claudiu Fierăscu, Alina Răditoiu, Cristian Andi Nicolae and Violeta Purcar

National Research and Development Institute for Chemistry and Petrochemistry—ICECHIM, 202 Splaiul Independentei, 6th District, 060021 Bucharest, Romania; monica.raduly@icechim.ro (F.M.R.); fierascu.radu@icechim.ro (R.C.F.); coloranti@icechim.ro (A.R.); ca_nicolae@yahoo.com (C.A.N.); violeta.purcar@icechim.ro (V.P.)
* Correspondence: vraditoiu@icechim.ro

Citation: Raduly, F.M.; Răditoiu, V.; Fierăscu, R.C.; Răditoiu, A.; Nicolae, C.A.; Purcar, V. Influence of Organic-Modified Inorganic Matrices on the Optical Properties of Palygorskite–Curcumin-Type Hybrid Materials. *Crystals* 2022, 12, 1005. https://doi.org/10.3390/cryst12071005

Academic Editor: Pier Carlo Ricci

Received: 1 June 2022
Accepted: 19 July 2022
Published: 20 July 2022

Publisher's Note: MDPI stays neutral with regard to jurisdictional claims in published maps and institutional affiliations.

Copyright: © 2022 by the authors. Licensee MDPI, Basel, Switzerland. This article is an open access article distributed under the terms and conditions of the Creative Commons Attribution (CC BY) license (https://creativecommons.org/licenses/by/4.0/).

Abstract: Clays are very important from an economic and application point of view, as they are suitable hosts for organic compounds. In order to diversify the fields of application, they are structurally modified by physical or chemical methods with cationic species, and/or different bifunctional compounds, such as organosilanes. In this study, palygorskite was modified with (3-Aminopropyl) triethoxysilane, which was subsequently modified at the amino group by grafting an acetate residue. By using this strategy, two types of host hybrid materials were obtained on which curcumin derivatives were deposited. The composites obtained were structurally characterized and their photophysical properties were investigated in relation to the structure of the host matrices and interactions with curcumin-type visiting species. The hybrid composites have different colors (orange, yellow, pink), depending on the polarity of the inorganic matrices modulated by different organic groups grafted at the surface. Fluorescence emission in the visible range is characterized by the presence of two emission maxima, one belonging to the chromophore and the other influenced by the physical interactions between auxochromes and host matrices. These hybrid materials, compared to other composite structures, are obtained by a simple adsorption process. They are temperature stable in aggressive environments (acid/base) and render the fluorescent properties of dyes redundant, with improved luminescent performance compared to them.

Keywords: hybrid materials; palygorskite; curcumin derivatives; X-ray diffraction; fluorescence emission

1. Introduction

Hybrid materials, consisting of carrier matrices of inorganic origin and an active compound of organic/inorganic type, are more and more common in current studies, especially in fields such as medicine, environmental protection or catalysis [1–5]. Inorganic matrices, such as clays, are very often used due to availability, low price, structural diversity and the possibility of recycling [6,7]. These are found as support matrices for two categories of materials, some that act as a collection vehicle and others as a transport and/or release of organic compounds. Of these, the former acts as an adsorbent and is usually used in water purification processes. In this case, the clays are conditioned by different methods so that the adsorption processes of organic/inorganic compounds that pollute the water are as efficient as possible [1,6]. These types of supports have been modified according to the nature of the active substances transported, the structure of clay-based matrices becoming much more complex. In this sense, the clays were modified with layer swelling compounds or organic substances with high affinity for bio-active compounds, so as to increase their adsorption capacity [7–10]. In some studies, palygorskite (attapulgite, $(Mg, Al)_2Si_4O_{10}(OH)4(H_2O)_2$), a magnesium aluminum phyllosilicate with a fibrillar structure, in monoclinic crystalline systems [11–13], is preferred for such applications. At the same time, palygorskite is known to be one of the oldest clays used to make hybrid materials (Maya blue pigment) obtained by the adsorption of natural indigo [14,15].

Palygorskite is currently treated with cationic substances [16,17], hydroxylamine derivatives [18] or silanes [19–22] to promote the adsorption of naturally occurring or synthesized organic compounds. Most often, these are essential oils [23], dyes [24–26] or enzymes [27] loaded on the inorganic matrices, resulting in hybrid materials that preserve their original properties in different environments, in which factors, such as light, temperature, or pH, can degrade organic compounds. The processes of loading inorganic matrices with bioactive compounds take place through ion exchange processes, adsorption at microwaves or ultrasound-assisted processes [20,28,29]. Such hybrid materials usually have antioxidant or antimicrobial properties and are used in the food and pharmaceutical industry [30–33].

One of the most sought after natural sources of antioxidant compounds is turmeric (Curcuma longa), whose main compounds are curcumin, demethoxy and bisdemethoxy derivatives. These phenolic compounds are found in studies that confirm their antioxidant, anti-inflammatory or antimicrobial properties [34–37]. Given these properties, we are currently trying to develop new and modern applications in areas related to human health. Thus, through various methods of conditioning with nanoparticles, natural polymers or clays, curcumin derivatives have found applications in the form of hybrid materials, in areas, such as food packaging, food additives, sensors, bioactive coatings and skin cosmetic products [38].

The aim of this study was to obtain hybrid materials made by depositing curcumin (CC) and a curcumin derivative (CCN) obtained by synthesis, on two types of silane-modified palygorskite. The hybrid materials obtained were structurally analyzed and the optical properties were evaluated. The influence of the structure of the modified inorganic support with different organic molecules having different polarities on the optical properties of the curcuminoids deposited on them was analyzed.

2. Materials and Methods

2.1. Materials

2.1.1. The Host Inorganic Matrix

The Palygorskite (PAL) clay used in this study was obtained from SERVA (Germany) with the chemical compositions presented in Table 1 (by X-ray fluorescence, XRF). The clay was activated in hydrochloric acid (5 N, HCl, Chimreactiv, Bucharest, Romania) and modified with (3-Aminopropyl)triethoxysilane (99%, APTES, Aldrich, St. Louis, MO, USA) and sodium chloroacetate 98%, ACD, Aldrich, St. Louis, MO, USA) using ethanol (96%, EtOH, Chimreactiv, Bucharest, Romania) as solvent.

Table 1. Elemental analysis of the palygorskite raw.

Element	Si	Al	Fe	Mg	K	Ca	Ti	L.E.
Raw (wt%)	25.79 (±0.48)	3.65 (±0.21)	4.19 (±0.09)	3.7 (±1.2)	0.48 (±0.02)	4.02 (±0.07)	0.63 (±0.05)	56.74 (±0.9)

L.E.—light elements (not detectable by XRF).

2.1.2. Curcumin Dyes

The curcumin dye (Figure 1) was synthesized at microwaves and purified using a method already published by our group [39]. Briefly, a mixture of boron trioxide (4 mmol), acetylacetone (8 mmol) and tributyl borate (3.2 mmol) was introduced in a porcelain capsule and was irradiated in a microwave oven at 300 W for 10 min. After the formation of the boron complex of acetylacetone, vanillin or 4-N,N-diethyl-benzaldehyde (7 mmol) and dodecylamine (0.162 mmol) were added, and the mixture was irradiated for another 20 min at 100 W. An aqueous solution of acetic acid (10% by weight) was added to the reaction mixture, the obtained suspension was filtered off, and the solid product was washed with cold water, and then dried. The obtained product was purified by recrystallization from a mixture of ethyl-acetate:methanol = 3:2 (v/v).

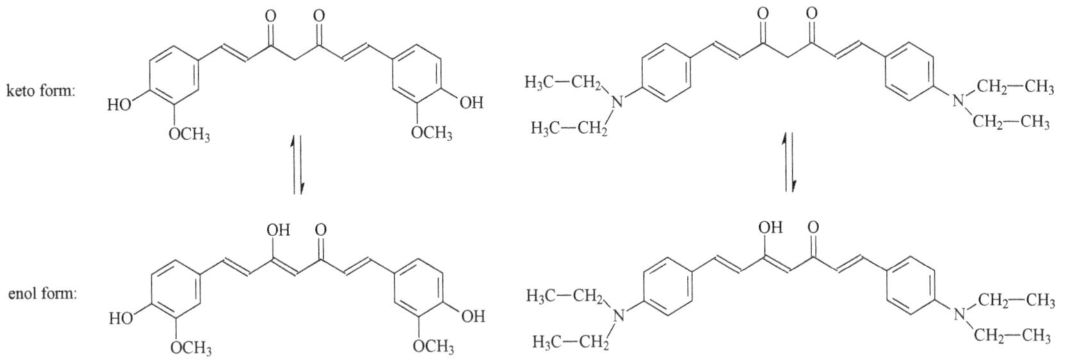

Figure 1. Keto-enol tautomeric forms of curcumin derivatives.

CC : (1E,6E)-1,7-bis(4-hydroxy-3-methoxyphenyl)-1,6-heptadiene-3,5-dione

CCN : (1E,6E)-1,7-bis(4- N, N-diethylphenyl)-1,6-heptadiene-3,5-dione

2.2. Methods

2.2.1. Obtaining the Palygorskite Functionalization

The clay was activated in a 5 M HCl solution at 100 °C for 4 h. The gelled composition was centrifuged, filtered, and then washed with water until neutral pH. Afterwards, the crude product was dried at 110 °C for 5 h in order to obtain activated PALs, which was dispersed into anhydrous ethanol, and the silane (APTES) was added to the mixture (1:1 mass/volume ratio) with vigorous stirring for 2 h at 45 °C. The product was filtered, washed with ethanol and dried at 110 °C. The modified clay with APTES (S1) was further treated with the sodium salt of chloroacetic acid (1:1 mass ratio) at 70 °C, for 4 h, then product (S2) was filtered, washed with ethanol and dried at 110 °C.

2.2.2. Obtaining the Colored Hybrid Materials

Onto every modified substrate (S1, S2) the curcuminoid dye was deposited (CC or CCN), obtaining hybrid materials. Thus, 1 g of clay was suspended in 50 mL dye alcoholic solution 0.54 g/L for 4 h at room temperature. It was the optimal concentration used to obtain colored composite materials whose optical properties do not vary significantly, compared to the use of higher concentrations of dye. The dispersion was filtered and the composite material dried at 105 °C.

2.2.3. Characterization

The elemental analysis of the material was performed using a Vanta C series handheld XRF (Olympus, Waltham, MA, USA), equipped with 40 kV X-ray tube with rhodium anode, Silicon Drift Detector, in the pre-calibrated GeoChem mode, and acquisition time 60 s for each beam. The equipment uses two different energy beams for the quantification of the elements: beam 2 (10 kV) for light elements (Mg, Al, Si, P, S, K, Ca, Ti, Mn) and beam 1 (40 kV) for the rest of the detectable elements.

XRD analyses were performed using a Rigaku SmartLab equipment (Rigaku Corporation, Tokyo, Japan), in $2\theta/\theta$ configuration, XRD diffractograms being recorded in the 2θ range 2–90°. The collected data were interpreted using the PDXL software (ver. 2.7.2.0, Rigaku Corporation, Tokyo, Japan), provided by Rigaku, and the identification of the phases present was performed by comparison with the ICDD (International Centre for Diffraction Data) database entries.

The structural differences of the host matrices were analyzed by FTIR measurements, recorded with a JASCO FT-IR 6300 instrument (Jasco Int. Co. Ltd., Tokyo, Japan), equipped with a Specac ATR Golden Gate (Specac Ltd., Orpington, UK) with KRS5 lens, in the range of 400 to 4000 cm^{-1} (128 accumulations at a resolution of 4 cm^{-1}).

Characterization of porous materials by surface area, pore volume and size measurements was performed using Nova 2200e Quantachrome (Quantachrome Instruments Corporate Drive, Boynton Beach, FL, USA) automated gas adsorption system. All the samples were out gassed at 110 °C for 4 h, after N$_2$ adsorption–desorption isotherms were measured at -196 °C. The specific surface areas (S_{BET}) and total pore volumes (V_{total}) of the samples were determined by BET (Brunauer–Emmett–Teller) and BJH (Barrett–Joyner–Halenda) methods, respectively.

The obtained results were interpreted and graphically presented using OriginPro 2018 data analysis software (ver. 9.50, OriginLab corporation, Northampton, MA, USA).

Thermogravimetric analysis (TGA) of the hybrid materials obtained was performed with a TGA Q5000IR instrument (TA Instruments, New Castle, DE, USA). The 5–9 mg samples were analyzed in platinum pans under the following conditions: heating ramp 10 °C/min up to 700 °C, Nitrogen 5.0 (99.999%), used as purge gas at a 50 mL/min flow rate.

Total color differences in CIELAB system, using a 10° standard observer and illuminant D65, and diffuse reflectance spectra, were measured with a V570 UV-VIS-NIR (Jasco Int. Co. Ltd., Tokyo, Japan) spectrophotometer equipped with a JASCO ILN-472 (150 mm) integrating sphere, using Spectralon as reference.

Evaluation of the fluorescence properties of hybrid materials was analyzed by recording steady-state fluorescence spectra on a JASCO FP 6500 spectrofluorimeter (Jasco Int. Co. Ltd., Tokyo, Japan) at 25 °C.

2.2.4. Evaluation of Resistance to Low Temperature, Acid, Alkali and Saline Solution

To evaluate the resistance of the hybrid materials to low temperature, the samples used to measure the specific surface area (cooled with liquid nitrogen, approximately 9 h) were investigated by recording their diffuse reflectance spectra using a UV-Vis spectrophotometer. For the other tests, 0.5 g of each hybrid material was placed in a paper cartridge and immersed in 25 mL 1.0 mol/L HCl or 1.0 mol/L NaOH or 1.0 mol/L NaCl solutions at room temperature for 24 h. The cartridges were then dried at 110 °C, and the diffuse reflectance spectra were measured for each hybrid material.

3. Results and Discussion

Following the adsorption process of curcumin derivatives on the inorganic matrix, decorated only with amino groups or with glycine residues, four types of hybrid materials of palygorskite-curcumin, were obtained. The degree of dye sorption on the two types of modified inorganic matrices was calculated using formula (1):

$$\text{Sorption degree (\%)} = [(C_0 - C_1) \cdot 100]/C_0 \tag{1}$$

where C_0, represent the initial and C_1 final concentrations (mg/L), respectively.

Depending on the design of the fibrous surface of the attapulgite clay, the optical properties of the hybrid materials obtained by the deposition of curcumin derivatives were investigated. For this purpose, the clay was decorated with amino groups, by treatment with (3-aminopropyl)triethoxysilanes. The treatment of S1 with sodium chloroacetate was performed in order to graft amphionic groups onto the structure of palygorskite (S2). Thus, the creation of the amphoteric character of the inorganic matrix aimed at improving the adsorption capacity of curcumin derivatives (Figure 2).

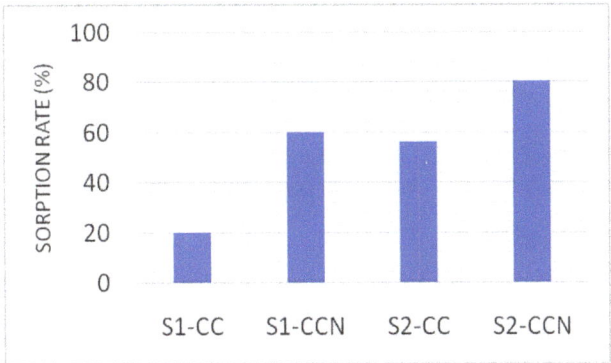

Figure 2. The sorption rate of curcumin dyes on palygorskite type hosting matrices.

3.1. Analytical Characterization of the Materials

3.1.1. XRD Analysis of Hybrid Materials

The X-ray diffractograms of the modified palygorskite, with organic molecules and composite materials obtained by depositing curcumin derivatives on the modified clays, are presented in Figures 3 and 4, respectively. The identification peaks corresponding to the palygorskite structure (ICCD 21-0958) are presented in the Supplementary Materials in the Table S1, as observed in other studies [7,9].

In the XRD pattern of palygorskite (PAL), the peak characteristic of the stratified structure can be found at a value of $2\theta = 8.58°$. By reacting with APTS, the substrate is silanized and functionalized with aminopropyl groups (S1). These groups intercalate in the interstitial space and cause a displacement of the peak at a value of $2\theta = 8.38°$. The reaction of the amino group with sodium chloroacetate leads to functionalization with a glycine residue (S2), in this case, the displacement of the peak taking place up to the value $2\theta = 8.20°$. The disappearance of this peak in the case of adsorption of CC and CCN dyes, which have larger molecules, indicates exfoliation, confirmed by the appearance of a wide band characteristic of amorphous phase, centered at $2\theta = 22°$.

This case is supported by the measurements of the curcumin molecule size, which is situated around the values 0.89–1.16 nm, depending on the conformation (cis-cis, cis-trans or trans-trans), with the highest most probable size of 1.05 nm, as already reported [9,40]. The values are around the distance $d = 1.04$ nm obtained from the characteristic (110) reflection of S1 and $d = 1.02$ nm in the case of S2.

Upon grafting, the X-ray diffractograms suffer some alterations (Figure 3), the most visible being for the APTES grafting, the decrease in intensity of peaks at 39.46° and 60.56° (2θ), peaks associated with the presence of quartz and, to a lesser extent, to those at 29.44°, 30.86° (2θ), the increase in intensity of the peaks at 21.32° and 25.72° (2θ), as well as the appearance of a supplemental peak at 22.28° (2θ), most probably due to an increase in the relative concentration of some minor phase upon purification. After the treatment of the APTES grafted sample with sodium chloroacetate, some minor changes in the X-ray diffractogram can be noticed, respectively, the increase in intensity of peaks at 8.58°, 27.52°, 57.34°, as well as the appearance of supplementary peaks at 31.75°, 33.18°, 45.56° and 56.62° (2θ), which could be assigned to the presence of acetate group.

After, the deposition of curcumin on the obtained samples, some changes can be observed on the X-ray diffractograms. Figure 4 presents the X-ray diffractograms obtained for the deposition of curcumin (S1-CC) and curcumin derivative (S1-CCN) on modified clay, respectively (Figure 4a). The major changes observed are the disappearance of some minor peaks associated with the palygorskite phase (i.e., at 2θ 29.44, 30.86, 57.34°), associated with the increase/apparition of some peaks corresponding to the curcumin phase: (S1-CC)-13.82°-(1,1,0), 20.83°-(2,1,−2), 26.63°-(5,1,−1), 36.48°-(6,1,−3), 50.13°-(6,3,−2), 81.54°-(5,2,9), (S1-CCN)–13.82°-(1,1,0), 17.77°-(4,0,0), 20.83°-(2,1,−2), 25.70°-(5,1,−1),

26.63°-(5,1,−1), 36.48°-(6,1,−3), 50.13°-(6,3,−2), 81.54°-(5,2,9). Major differences are recorded upon the deposition of dye on support S3 (Figure 4b). It can be observed that the intensity of some peaks associated with the presence of curcumin has increased, i.e., at 26.62°-(2,1,3), 60.09°-(10,0,−6), 68.25°-(5,0,−9), 79.93°-(1,6,0) and 81.46°-(17,1,0). However, the major change visible from the diffractogram is the apparition of a specific broad "hump", centered around 22° (2 θ), which suggests the apparition of an amorphous phase at the surface of the material, following the interaction curcumin/support material.

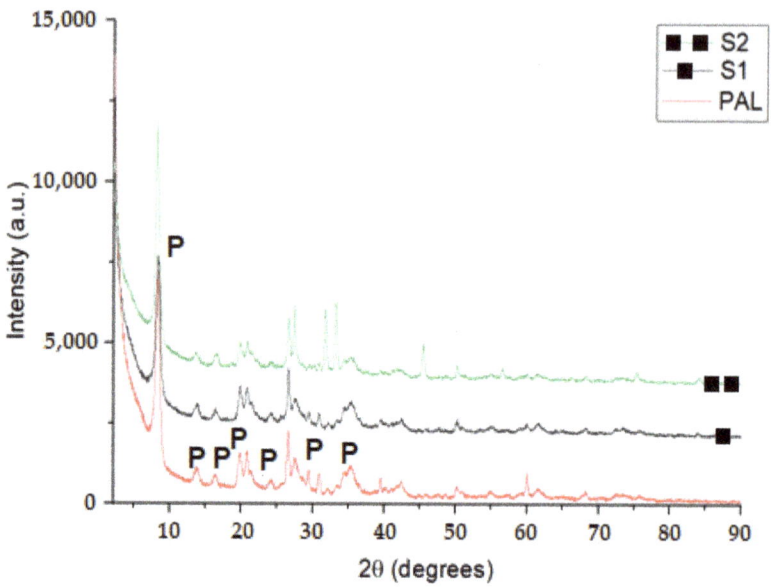

Figure 3. XRD patterns of the support materials.

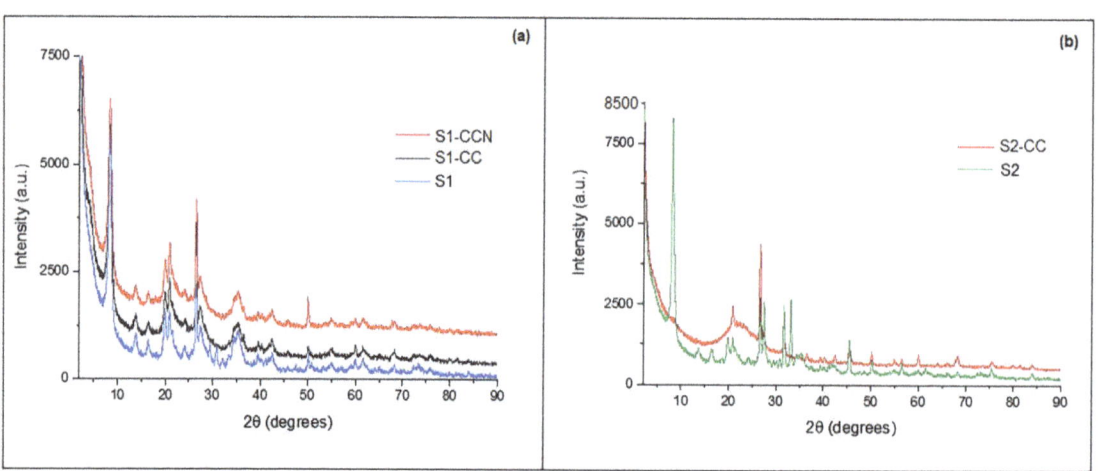

Figure 4. XRD patterns of the hybrid materials modified with aminopropyl groups (**a**) and glycine residue (**b**).

3.1.2. N2 Adsorption Isotherms

Nitrogen adsorption–desorption isotherms of modified clay and colored nanocomposites at 77 K are type II in concordance IUPAC classification. These isotherms show hysteresis with large loops, characteristic to mesoporous materials, thus confirming the textural change of the clay. The treatment of clay with APTES, leads to the decrease in the specific surface and the formation of meso- and macroporous surfaces, with an average pore diameter of 30 nm. After the modification of the composite with sodium chloroacetate, the characteristics mentioned above become more pronounced. Thus, the hybrid materials obtained have an even smaller specific surface area, but the pore volume increases (Table 2). The modification of the amino groups with acetate groups led to the increase in the interstitial volume in the structure of the palygorskite–APTES hybrid, the pore diameter obtained being in the range of 10–50 nm (Figure 5). The process of adsorbing the dye on the modified clay is influenced by the texture of the substrate, but at the same time it depends a lot on the chemical structure of the dye.

This is evident in the deposition of curcumin on the two types of hybrid materials, through which colored composites were obtained with a smaller pore volume than the S1 hybrid support, where we assume that the adsorption of the dye took place mainly in mesopores of the support structure. The adsorption of the CCN dye, which has in its chemical structure voluminous N-diethyl amine groups, leads to composite materials with a higher pore volume than that of the support. These results were most likely obtained by the partial adsorption of the dye in the structure of the hybrid material, confirmed by the increase in the specific surface area at 97 m^2/g. In the case of S2 support, the chemical and textural structure increases the affinity for curcumin derivatives. The adsorption process is favored by the amphoteric character of the support, and the dyes interact more intensely with it through the physical bonds established with the acetate and amino groups grafted on the surface. The presence of smaller pores of 3–5 nm in the structure of the S2–CCN hybrid material is probably due to the aggregation phenomena between the dye molecules. The structural aggregates of the dyes, deposited on the S2 support, lead to amorphous structures, confirmed by XRD analyzes.

Table 2. Specific surface area, pore volume and average pore diameter of PAL and organic–inorganic hybrids.

Sample	S_{BET} (m^2/g)	V_{total} (cm^3/g)	Dmed$_{pore}$ (nm)
PAL	117	0.327	13
S1	86	0.281	30
S1-CC	81	0.193	10
S1-CCN	97	0.298	13
S2	36	0.319	15
S2-CC	40	0.167	17
S2-CCN	24	0.272	3

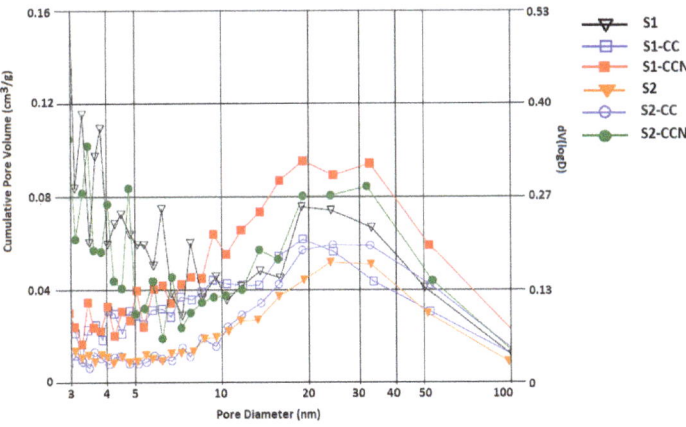

Figure 5. Pore-size distribution of the hybrid materials.

3.1.3. Characterization of Hybrid Materials by ATR-FTIR Spectra

Figure 6 shows the FTIR spectra of organically modified clays and colored composites. The range 1800–1300 cm^{-1} is representative for the identification of the vibration bands of the bonds in the organic compounds grafted onto the hybrid materials. Peaks characteristic to inorganic matrices in this interval are found around values of 1650 cm^{-1}, this being the characteristic to the vibration band of the bending and stretching of water molecules. The spectra for the modified palygorskite includes the bands at 3610 and 3290 cm^{-1}, corresponding to the stretching vibrations of the OH groups related to Al, Mg or Si of the network. The band in the range 1050–914 cm^{-1} was assigned to the Si-O stretching vibration, respectively, and the band at 796 cm^{-1}, to the Si-O-Si vibration.

The presence of organic molecules on the inorganic structure is confirmed by the bands from 1560 cm^{-1}, 1490 cm^{-1} and 1430 cm^{-1} attributed to the stretching vibration of the –NH$_2$ and C–N bonds of APTES, respectively. The wide band at 1650 cm^{-1} is displaced and is attributed to the C = C bonds in the chemical structure of the dye that overlaps with the vibration band of the water molecules in the inorganic network (Figure 6a). Curcumin-colored hybrid materials (S1-CC), due to the low adsorption of the dye, are characterized by low intensities of the characteristic vibration bands. On the other hand, the molecular structure of the CCN dye (in the Supplementary Materials, Figure S1) deposited on the S1 support contributes to the intensification of the vibration bands characteristic of amines. At the same time, the appearance of new peaks at 1590 cm^{-1} and 1520 cm^{-1} attributed to secondary amines, respectively, means the NH band confirms the interactions between the dye and the modified clay.

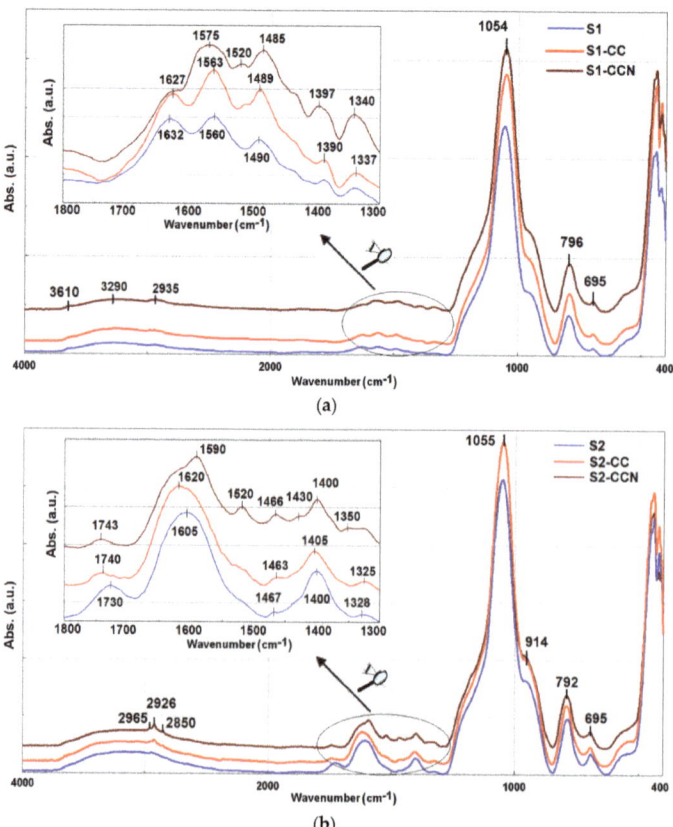

Figure 6. FTIR spectra of the hybrid materials modified with aminopropyl groups (**a**) and glycine residue (**b**).

For hybrid materials treated with sodium chloroacetate, the molecular structures formed are more complex. The FTIR spectra confirm the grafting of the carboxyl groups by the presence of bands from 1740 cm^{-1} attributed to the stretching vibration C = O. Due to the multiple interactions between atoms of the amino groups, acetate and hydroxy, their characteristic peaks are overlapped. Thus, the hybrid materials are characterized by the presence of wide bands with obvious displacements of the absorption maxima for the colored composites compared to the modified clays (Figure 6b).

3.1.4. Thermogravimetric Analysis of Composite Materials

To evaluate the thermal stability and decomposition of hybrid materials, TG curves were recorded for the inorganic matrix modified with organic groups, and then loaded with curcumin dyes (Figure 7). At the same time, the decomposition temperatures and mass losses characteristic to the analyzed hybrid materials are presented in Table 3.

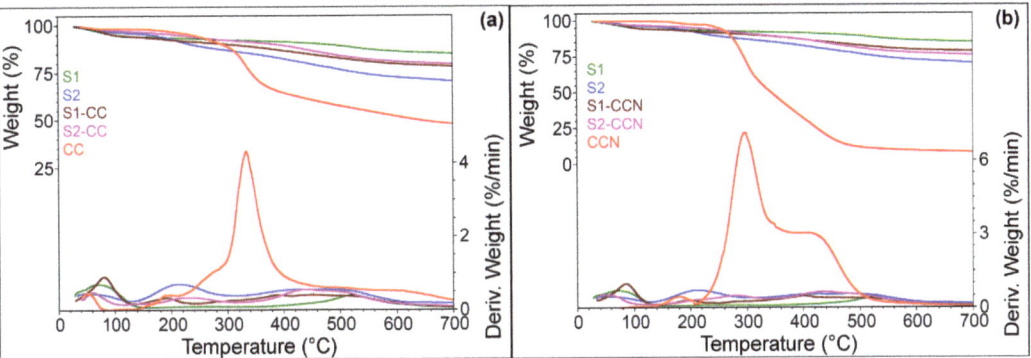

Figure 7. Thermogravimetric curves of hybrid materials modified with aminopropyl groups and CC dye (a) and hybrid materials modified with glycine residue and CCN dye (b).

Table 3. Characteristic decomposition temperatures and mass losses of hybrid materials and curcumin dyes.

Sample	RT–140 °C Wt. Loss (%)	140–285 °C Wt. Loss (%)	140–285 °C Tmax (°C)	285–410 °C Wt. Loss (%)	285–410 °C Tmax (°C)	410–700 °C Wt. Loss (%)	410–700 °C Tmax (°C)	Residue 700 °C (%)
S1	5.89	1.74	248.8	7.05	-	-	521.10	85.27
S2	3.66	9.63	213.2	15.90	429.2	-	478.30	70.76
S1-CC	5.52	3.77	190.7	4.12		8.08	452.3	78.51
S2-CC	2.82	4.27	223.2	3.63		9.55	441.2	79.73
CC	1.53	27.11	271.5	15.96	333.0	7.35		48.04
S1-CCN	5.79	1.87	197.4	2.94	279.1	10.48	400.8	78.93
S2-CCN	3.34	1.90	202.1	5.11	281.7	13.46	436.3	76.18
CCN	0.23	2.44	179.7	46.07	296.9	42.86	405.1	8.39

By comparing the TG curves of composite materials, there can be observed a first stage of constant mass loss around 5.5% (±0.3) for modified materials with APTES (with and without dye), and 3.3% (±0.3) for those modified with glycine residue, respectively. This process takes place in the temperature range of 50–140 °C and can be attributed to the desorption of water adsorbed on the surface and of zeolitic water, existing in the channels of the inorganic matrix. The second stage of thermal decomposition, which involves mass loss, takes place differently and depends on the molecular structure of the dye loaded on the support. It can also be seen that the mass loss differs for the two types of modified substrate. For S2, the thermal effects are more dynamic, reaching a maximum temperature of 213 °C. For this temperature range, the mass loss for S2, around 9.6% is attributed to the decomposition of the organic residues grafted on PAL, a process that continues up to 429 °C, the stage in which the mass loss is 15.9%. The mass loss, is almost double for S2, compared to S1 and is due to the additional load on the inorganic matrix with the rest of glycine, confirmed by XRD and FTIR analyses.

However, when the two types of support matrix are loaded with CC, the mass losses are comparative (~4%). The maximum decomposition temperature at 119 °C for S1-CC can be attributed to the weak interactions between the dye and the host matrix. In comparison, it can be seen that the mass loss for S2-CC occurs at a higher temperature (223 °C) and can be attributed to the stronger interactions between the amphoteric support and the dye. For hybrid materials loaded with CCN, the thermal decomposition effects that take place in the second stage are similar and the mass losses are less than 2%. These thermal events continue in the next temperature range 285–410 °C. At this stage, S2-CCN has a higher mass loss than S1-CCN, motivated by the decomposition of the dye bound to the PAL surface modified with glycine residues, as confirmed by XRD and BET analysis. In the last temperature range of 410–700 °C, the composite materials have a mass loss between 8–13%. These thermal effects occur around the temperature of 400 °C and are attributed to the loss of water due to the aluminol and silanol groups condensation [9,10].

3.2. Photophysical Studies

3.2.1. Spectrophotometric Properties of Hybrid Materials

In order to evaluate the chromatic characteristics of the colored hybrid materials (Figure 8a), the color coordinates L*, a*, b* were compared. The results of total color differences in the CIELAB system of materials colored with curcumin dyes showed that the structure of the inorganic material greatly influences the final color of the composites. In S1–dye hybrids, dyes interact with the support on which amino groups are grafted, by predominantly polar bonds (CCN) and less by hydrogen bonds (CC).

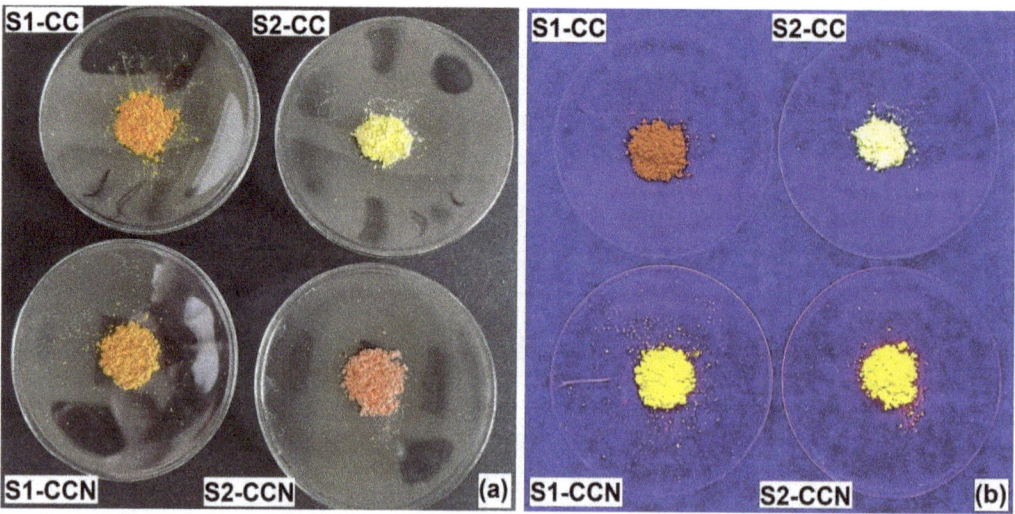

Figure 8. Images of hybrid materials exposed to daylight (**a**) and fluorescence emission during excitation at 365 nm (**b**).

The coordinates are comparable, the brightness L* being slightly lower and a* slightly higher for S1-CC (Figure 9a), this leading to a darker orange shade than S1-CCN (Figure 9c). For the inorganic support decorated with both amino and acetate groups, the parameters of colored hybrid materials differ significantly. The polarity of the inorganic support increases the polar interactions with the dye. In the case of the S2–CC type hybrid, the polar physical bonds between the dye and the support lead to a drastic decrease in its a* value and increase in the brightness, obtaining a yellow composite (Figure 9b). For the CCN composite, the presence of acetate groups has the effect of decreasing the value of b*, the shift to red having the effect of changing the color to pink (Figure 9d).

The resistance of colored composites to low temperature, alkaline, acidic and saline environments was assessed by color measurements for the exposed samples. By comparing the color parameters, it can be seen that the hybrid materials have good stability at low temperatures, and in the case of S1-CC, a more pronounced hypsochromic process takes place. In an alkaline environment, the bonds between the dye and the substrate are affected, which leads to the release of the dye from the surface of the modified clay. In acidic environments, the hybrid materials show stability, but with a color change to yellow for S1-CC and red for S2-CCN, respectively, most likely determined by the tautomeric structures of the dye. In the saline environment, the composite materials did not have essential changes, except for S2-CCN, which had a bathochromic displacement. However, the evaluation of resistance is done by calculating the total color differences, which in turn can be broken down into components related to the influence of hue, chroma and lightness, as it can be seen in Table S2 (in the Supplementary Materials). The stability of the colored materials was evaluated by measuring color differences (ΔE*) between the sample before and after performing tests on various environmental factors. It was observed by comparing the total color differences, ΔE*, that the smallest variations were recorded for the S2-CCN sample, which is further evidence of the strong interactions established between the curcumin derivative and the host matrix.

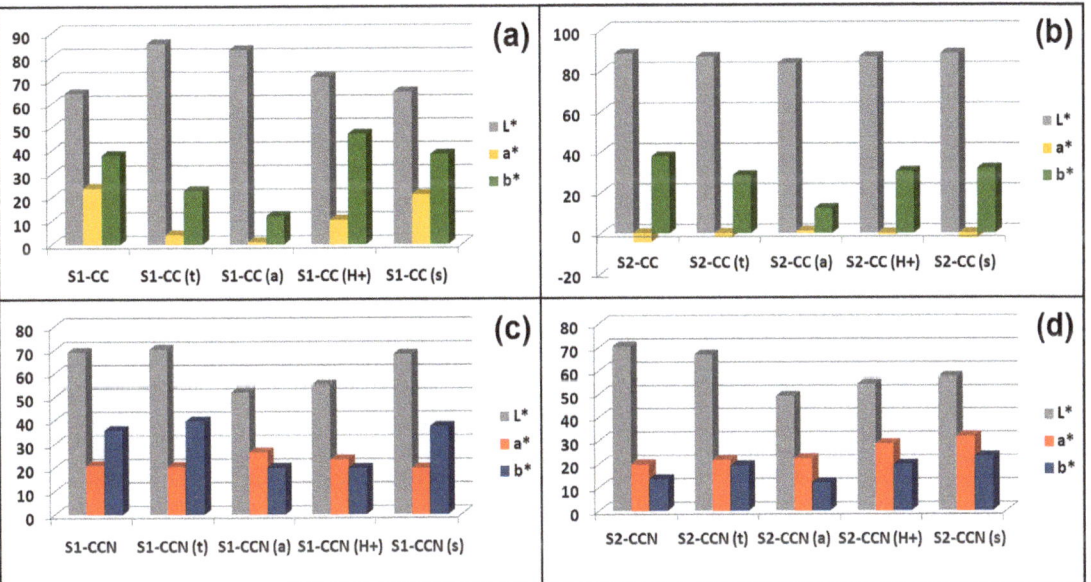

Figure 9. Color parameters in the CIELAB system for colored nanocomposites (S1-CC (**a**), S2-CC (**b**), S1-CCN (**c**), S2-CCN (**d**)) and after testing their resistance to low temperature (t), alkaline (a), acid (H+) and saline (s) solutions.

3.2.2. Fluorescent Properties of Hybrid Materials

The fluorescent properties of the colored hybrid materials are shown in Figure 8b. All spectra of hybrid materials and solid dyes, respectively, show two emission maxima (Figure 10). The first maximum of fluorescence emission at 480 nm is characteristic to the chromophore and is common to the two dyes. The second peak is determined by the auxochromes attached to the chromophore. Thus, by attaching the hydroxy and methoxy groups (CC), a maximum emission is obtained at 590 nm, with an increased intensity compared to the peak at 480 nm. The attachment of amino groups (CCN) on aromatic rings has the effect of extending the conjugation and moving the second maximum of fluorescence emission to a longer wavelength (685 nm). The explanation for the allure of the fluorescence emission spectra is related to the presence of the basic aminopropyl groups in S1, respectively, the amphionic groups of the glycine residue in S2, groups that interact differently with the terminal groups of curcuminoid dyes, phenolic for CC and dialkylarylamine for CCN. All these interactions between the dye and the host matrix lead to the increased fluorescence intensity of hybrid materials compared to that of the curcumin dyes (Figure 10). From the variation of the basic/acid character of the four groups, there are interactions through hydrogen bonds that can lead as far as the formation of ions, as in the case of the S2-CCN system, which has a very intense emission band displaced hypsochromic at 580 nm. At the same time, for the S2-CC system, the same emission band is found at 515 nm and the hypsochromic change can be attributed to the dye that acts as a proton acceptor [41]. Interactions, such as hydrogen bonds between the dye and the host matrix, also affect the chromophore and a hypsochromic shift at 425 nm is observed. In the case of CCN, only the intensities of these bands are influenced due to the surrounding environment of the matrices. In accordance with the results obtained for the color parameters, the S1–CC hybrid, in whose structure hydrogen bonds are established, in addition to the polar interactions, has the lowest intensity of fluorescence emission. By establishing hydrogen bonds, a strong path for quenching fluorescence emission is created, thus explaining the noted effect.

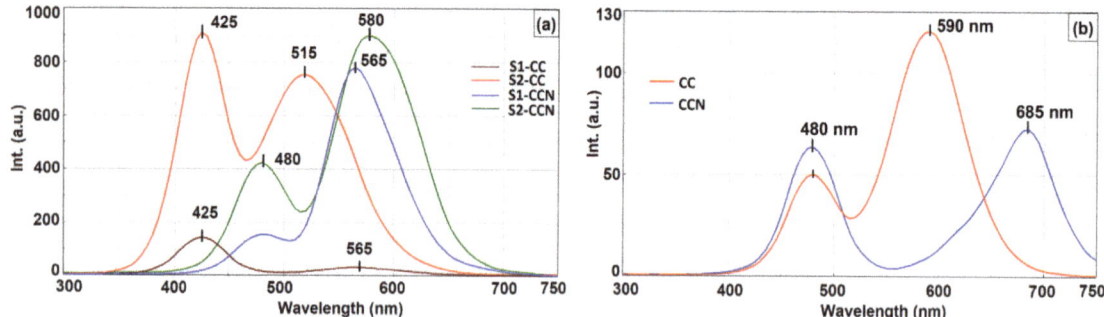

Figure 10. Fluorescence emission spectra of colored nanocomposites (**a**) and curcumin derivative (**b**).

In the case of hybrids loaded with CCN, the maximum emission characteristic to the curcumin chromophore is unchanged at 480 nm, while the polar interactions between the chromophore and the support determined a peak at 565 nm for the modified support with APTES. The strongly blue shift of 120 nm is probably due to the steric effects that modify the planarity of the dye molecules and the extended interaction with the support. Increasing the polarity of the inorganic matrix by grafting acetate groups leads to an increase in fluorescence intensity for the entire composite system. In particular, the addition of anionic groups produces, besides the intensification of the fluorescence emission, a 105 nm hypsochromic shift of the long wavelength emission band compared to that of the bulk dye.

4. Conclusions

In this study, the structural modifications of palygorskite with organic groups were made in order to obtain hybrid materials compatible with curcumin-derived dyes. The morphostructural characteristics and optical properties of the colored hybrid materials were studied in relation to the design of modified clay hosting material. The adsorption capacity of two curcuminoid dyes on two types of hybrid matrices derived from palygorskite were studied, noting that the degree of sorption is higher in the case of dye with basic peripheral groups and amphionic modifier grafted onto the palygorskite network. According to XRD data, the introduction of curcuminoid dyes into the palygorskite network leads to exfoliation, in the case of the host network modified with glycine residues. The presence of acid or basic groups grafted onto the surface of palygorskite that interact differently with the terminal residues of curcuminoid dyes leads to large Stokes shifts. Following the characterization–structure characteristics, dye-loaded composites can be used in various fields, such as sensor materials.

Supplementary Materials: The following supporting information can be downloaded at: https://www.mdpi.com/article/10.3390/cryst12071005/s1, Figure S1: FTIR spectra of the curcumin dyes [39]; Table S1: The XRD peaks corresponding to the palygorskite structure; Table S2: Color differences for hybrid colored materials, exposed to various environmental factors.

Author Contributions: Conceptualization, F.M.R. and V.R.; methodology, A.R. and V.P.; software, R.C.F.; validation, F.M.R. and A.R.; formal analysis, V.P., C.A.N. and R.C.F.; investigation, F.M.R., A.R. and V.R.; writing—original draft preparation, F.M.R. and V.P.; writing—review and editing, R.C.F. and A.R.; visualization, R.C.F.; supervision, V.R.; project administration, F.M.R., A.R. and R.C.F. All authors have read and agreed to the published version of the manuscript.

Funding: This work was supported by a grant of the Ministry of Research, Innovation and Digitization, CNCS/CCCDI—UEFISCDI, project number PN-III-P2-2.1-PED-2019-1471, within PNCDI III and by the INCDCP ICECHIM Bucharest 2019–2022 Core Program PN. 19.23–Chem-Ergent, Project No. 19.23.03.04. This work was supported by the Ministry of Research, Innovation and Digitization through Program 1-Development of the national research and development system, Subprogram 1.2 -Institutional performance- Projects to finance excellence in RDI, Contract no. 15 PFE/2021.

Data Availability Statement: Not applicable.

Conflicts of Interest: The authors declare no conflict of interest.

References

1. Gil, A.; Santamaría, L.; Korili, S.A.; Barbosa, L.V.; de Souza, S.D.; Marçal, L.; de Faria, E.H.; Ciuffi, K.J. A review of organic-inorganic hybrid clay based adsorbents for contaminants removal: Synthesis, perspectives and applications. *J. Environ. Chem. Eng.* **2021**, *9*, 105808. [CrossRef]
2. Kim, M.H.; Choi, G.; Elzatahry, A.; Vinu, A.; Choy, Y.B.; Choy, J.H. Review of Clay-Drug Hybrid Materials for Biomedical Applications: Administration Routes. *Clays Clay Miner.* **2016**, *64*, 115–130. [CrossRef] [PubMed]
3. da Rocha, M.C.; Galdino, T.; Trigueiro, P.; Honorio, L.M.C.; de Melo Barbosa, R.; Carrasco, S.M.; Silva-Filho, E.C.; Osajima, J.A.; Viseras, C. Clays as Vehicles for Drug Photostability. *Pharmaceutics* **2022**, *14*, 796. [CrossRef]
4. Charisiou, N.D.; Sebastian, V.; Hinder, S.J.; Baker, M.A.; Polychronopoulou, K.; Goula, M.A. Ni Catalysts Based on Attapulgite for Hydrogen Production through the Glycerol Steam Reforming Reaction. *Catalysts* **2019**, *9*, 650. [CrossRef]
5. Ispas, G.C.; Manea, R.; Brazdis, R.I.; Baroi, A.M.; Fistos, T.; Fierascu, R.C.; Raduly, M.F. Iron Oxide/Phosphatic Materials Composites with Potential Applications in Environmental Protection. *Materials* **2020**, *13*, 5034. [CrossRef]
6. Wang, Y.; Feng, Y.; Jiang, J.; Yao, J. Designing of Recyclable Attapulgite for Wastewater Treatments: A Review. *ACS Sustain. Chem. Eng.* **2019**, *7*, 1855–1869. [CrossRef]
7. Silva, V.C.; Araújo, M.E.B.; Rodrigues, A.M.; Cartaxo, J.M.; Menezes, R.R.; Neves, G.A. Adsorption Behavior of Acid-Treated Brazilian Palygorskite for Cationic and Anionic Dyes Removal from the Water. *Sustainability* **2021**, *13*, 3954. [CrossRef]
8. Zhang, Y.; Wang, W.; Mu, B.; Wang, Q.; Wang, A. Effect of grinding time on fabricating a stable methylene blue/palygorskite hybrid nanocomposite. *Powder Technol.* **2015**, *280*, 173–179. [CrossRef]
9. Zhuang, G.; Li, L.; Li, M.; Yuan, P. Influences of micropores and water molecules in the palygorskite structure on the color and stability of Maya blue pigment. *Microporous Mesoporous Mater.* **2022**, *330*, 111615. [CrossRef]
10. Sousa, M.U.; Rodrigues, A.M.; Araujo, M.E.B.; Menezes, R.R.; Neves, G.A.; Lira, H.L. Adsorption of Sodium Diclofenac in Functionalized Palygoskite Clays. *Materials* **2022**, *15*, 2708. [CrossRef]
11. Giustetto, R.; Chiari, G. Crystal structure refinement of palygorskite from neutron powder diffraction. *Eur. J. Mineral.* **2004**, *16*, 521–532. [CrossRef]
12. Benaissa, R.; Ezzouhra, Z.; Lahcen, B.; Omar, L.; Francis, M.; Lahcen, D.; Christine, L.M.; M'Barek, A.; François, S.; Amane, J.; et al. Comprehensive physicochemical study of dioctahedral palygorskite rich clay from Marrakech High Atlas (Morocco). *Phys. Chem. Miner.* **2013**, *40*, 411–424. [CrossRef]
13. Almeida, R.; Ferraz, E.; Santarén, J.; Gamelas, J.A.F. Comparison of Surface Properties of Sepiolite and Palygorskite: Surface Energy and Nanoroughness. *Nanomaterials* **2021**, *11*, 1579. [CrossRef] [PubMed]

14. Straulino-Mainou, L.; Pi-Puig, T.; Lailson-Tinoco, B.; Castro-Chong, K.; Urbina-Lemus, M.F.; Escalante-Gonzalbo, P.; Sedov, S.; Flores-Morán, A. Maya Blue Used in Wall Paintings in Mexican Colonial Convents of the XVI Century. *Coatings* **2021**, *11*, 88. [CrossRef]
15. Grazia, C.; Buti, D.; Amat, A.; Rosi, F.; Romani, A.; Domenici, D.; Sgamellotti, A.; Miliani, C. Shades of blue: Non-invasive spectroscopic investigations of Maya blue pigments. From laboratory mock-ups to Mesoamerican codices. *Herit. Sci.* **2020**, *8*, 1. [CrossRef]
16. Guo, H.; Xia, K.; Cao, M.; Zhang, X. Surface Modification of Attapulgite by Grafting Cationic Polymers for Treating Dye Wastewaters. *Materials* **2021**, *14*, 792. [CrossRef] [PubMed]
17. Wang, S.; Shen, Y.; Chen, X.; Dong, L.; Bao, H.Y.M.; Li, Y. Cationic surfactant-modified palygorskite particles as effective stabilizer for Pickering emulsion gel formation. *Appl. Clay Sci.* **2022**, *219*, 106439. [CrossRef]
18. Zhang, Z.; Wang, W.; Kang, Y.; Wang, Q.; Wang, A. Structure evolution of brick-red palygorskite induced by hydroxylammonium chloride. *Powder Technol.* **2018**, *327*, 246–254. [CrossRef]
19. Kasprzhitskii, A.; Lazorenko, G.; Kruglikov, A.; Kuchkina, I.; Gorodov, V. Effect of Silane Functionalization on Properties of Poly(Lactic Acid)/Palygorskite Nanocomposites. *Inorganics* **2021**, *9*, 3. [CrossRef]
20. da Silva, T.F.; de Souza, G.P.M.; de Melo Morgado, G.F.; Wearn, Y.N.; Albers, A.P.F.; Quinteiro, E.; Passador, F.R. A Brief Review of the Latest Advances of Attapulgite as a Reinforcing Agent in Polymer Matrix Nanocomposites. *Am. J. Eng. Appl. Sci.* **2021**, *14*, 292–307. [CrossRef]
21. Ferraz, E.; Alves, L.; Sanguino, P.; Santarén, J.; Rasteiro, M.G.; Gamelas, J.A.F. Stabilization of Palygorskite Aqueous Suspensions Using Bio-Based and Synthetic Polyelectrolytes. *Polymers* **2021**, *13*, 129. [CrossRef] [PubMed]
22. Abbassia, H.; Abidia, R.; Zayani, M.B. A short review on the silylated clays—Polymer nanocomposites: Synthesis, properties and applications. *J. Mar. Chim. Heterocycl.* **2021**, *20*, 117–134.
23. Zhong, H.; Mu, B.; Yan, P.; Jing, Y.; Hui, A.; Wang, A. A comparative study on surface/interface mechanism and antibacterial properties of different hybrid materials prepared with essential oils active ingredients and palygorskite. *Colloids Surf. A Physicochem. Eng. Asp.* **2021**, *618*, 126455. [CrossRef]
24. Raduly, M.; Raditoiu, V.; Raditoiu, A.; Wagner, L.; Amariutei, V.; Nicolae, C. Luminescent hybrid materials based on curcumin derivatives embedded in palygorskite. *Mat. Plast.* **2018**, *55*, 63–67. [CrossRef]
25. Wang, Q.; Mu, B.; Zhang, Y.; Zhang, J.; Wang, A. Palygorskite-based hybrid fluorescent pigment: Preparation, spectroscopic characterization and environmental stability. *Microporous Mesoporous Mater.* **2016**, *224*, 107–115. [CrossRef]
26. Silva, G.T.M.; Silva, C.P.; Gehlen, M.H.; Oake, J.; Bohne, C.; Quina, F.H. Organic/inorganic hybrid pigments from flavylium cations and palygorskite. *Appl. Clay Sci.* **2018**, *162*, 478–486. [CrossRef]
27. Zhou, P.; Tan, Y.Q.; Zhang, L.; Zhou, Y.M.; Gao, F.; Zhou, G.H. Effects of dietary supplementation with the combination of zeolite and attapulgite on growth performance, nutrient digestibility, secretion of digestive enzymes and intestinal health in broiler chickens. *Asian-Australas. J. Anim. Sci.* **2014**, *27*, 1311–1318. [CrossRef]
28. Suarez, M.; García-Rivas, J.; Morales, J.; Lorenzo, A.; García-Vicente, A.; García-Romero, E. Review and new data on the surface properties of palygorskite: A comparative study. *Appl. Clay Sci.* **2022**, *216*, 106311. [CrossRef]
29. Ruiz-Hitzky, E.; Ruiz-García, C.; Fernandes, F.M.; Lo Dico, G.; Lisuzzo, L.; Prevot, V.; Darder, M.; Aranda, P. Sepiolite-Hydrogels: Synthesis by Ultrasound Irradiation and Their Use for the Preparation of Functional Clay Based Nanoarchitectured Materials. *Front. Chem.* **2021**, *9*, 733105. [CrossRef]
30. Cheng, Y.; Xu, Q.; Chen, Y.; Su, Y.; Wen, C.; Zhou, Y. Modified Palygorskite Improves Immunity, Antioxidant Ability, Intestinal Morphology, and Barrier Function in Broiler Chickens Fed Naturally Contaminated Diet with Permitted Feed Concentrations of Fusarium Mycotoxins. *Toxins* **2018**, *10*, 482. [CrossRef]
31. Yang, W.; Chen, Y.; Cheng, Y.; Wen, C.; Zhou, Y. Effects of zinc bearing palygorskite supplementation on the growth performance, hepatic mineral content, and antioxidant status of broilers at early age. *Asian-Australas. J. Anim. Sci.* **2017**, *30*, 1006–1012. [CrossRef] [PubMed]
32. Yang, F.; Wang, A. Recent researches on antimicrobial nanocomposite and hybrid materials based on sepiolite and palygorskite. *Appl. Clay Sci.* **2022**, *219*, 106454. [CrossRef]
33. Hui, A.; Yang, F.; Yan, R.; Kang, Y.; Wang, A. Palygorskite-Based Organic–Inorganic Hybrid Nanocomposite for Enhanced Antibacterial Activities. *Nanomaterials* **2021**, *11*, 3230. [CrossRef] [PubMed]
34. Urošević, M.; Nikolić, L.; Gajić, I.; Nikolić, V.; Dinić, A.; Miljković, V. Curcumin: Biological Activities and Modern Pharmaceutical Forms. *Antibiotics* **2022**, *11*, 135. [CrossRef] [PubMed]
35. Rathore, S.; Mukim, M.; Sharma, P.; Devi, S.; Nagar, J.C.; Khalid, M. Curcumin: A Review for Health Benefits. *Int. J. Res. Rev.* **2020**, *7*, 273–290.
36. Pandey, A.; Chaturvedi, M.; Mishra, S.; Kumar, P.; Somvanshi, P.; Chaturvedi, R. Reductive metabolites of curcumin and their therapeutic effects. *Heliyon* **2020**, *6*, e05469. [CrossRef] [PubMed]
37. Oglah, M.K.; Mustafa, Y.F.; Bashir, M.K.; Jasim, M.H. Curcumin and Its Derivatives: A Review of Their Biological Activities. *Syst. Rev. Pharm.* **2020**, *11*, 472–481.
38. Raduly, F.M.; Raditoiu, V.; Raditoiu, A.; Purcar, V. Curcumin: Modern applications for a versatile additive. *Coatings* **2021**, *11*, 519. [CrossRef]

39. Raduly, M.F.; Raditoiu, V.; Raditoiu, A.; Wagner, L.E.; Amariutei, V.; Ailiesei Darvaru, G. Facile synthesis of curcumin and curcuminoid-like derivatives at microwaves. *Rev. Chim.* **2018**, *69*, 1327–1331. [CrossRef]
40. Ilnytskyi, J.; Patsahan, T.; Pizio, O. On the properties of the curcumin molecule in water. Exploration of the OPLS—United atom model by molecular dynamics computer simulation. *J. Mol. Liq.* **2016**, *223*, 707–715. [CrossRef]
41. Dearde, J.C. Investigation of the self-association of phenols and anilines by ultraviolet spectroscopy. *Can. J. Chem.* **1963**, *41*, 2683–2691. [CrossRef]

MDPI AG
Grosspeteranlage 5
4052 Basel
Switzerland
Tel.: +41 61 683 77 34

Crystals Editorial Office
E-mail: crystals@mdpi.com
www.mdpi.com/journal/crystals

Disclaimer/Publisher's Note: The statements, opinions and data contained in all publications are solely those of the individual author(s) and contributor(s) and not of MDPI and/or the editor(s). MDPI and/or the editor(s) disclaim responsibility for any injury to people or property resulting from any ideas, methods, instructions or products referred to in the content.

www.ingramcontent.com/pod-product-compliance
Lightning Source LLC
LaVergne TN
LVHW070628100526
838202LV00012B/755